NGSS FOR ALL STUDENTS

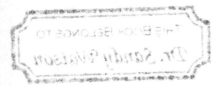

NGSS FOR ALL STUDENTS

OKHEE LEE
EMILY MILLER
RITA JANUSZYK

EDITORS

National Science Teachers Association
Arlington, Virginia

National Science Teachers Association

Claire Reinburg, Director
Wendy Rubin, Managing Editor
Andrew Cooke, Senior Editor
Amanda O'Brien, Associate Editor
Donna Yudkin, Book Acquisitions Coordinator

ART AND DESIGN
Will Thomas Jr., Director
Himabindu Bichali, Graphic Designer, cover and interior design

PRINTING AND PRODUCTION
Catherine Lorrain, Director

NATIONAL SCIENCE TEACHERS ASSOCIATION
David L. Evans, Executive Director
David Beacom, Publisher

1840 Wilson Blvd., Arlington, VA 22201
www.nsta.org/store
For customer service inquiries, please call 800-277-5300.

Copyright © 2015 by the National Science Teachers Association.
All rights reserved. Printed in the United States of America.
18 17 16 15   4 3 2 1

NSTA is committed to publishing material that promotes the best in inquiry-based science education. However, conditions of actual use may vary, and the safety procedures and practices described in this book are intended to serve only as a guide. Additional precautionary measures may be required. NSTA and the authors do not warrant or represent that the procedures and practices in this book meet any safety code or standard of federal, state, or local regulations. NSTA and the authors disclaim any liability for personal injury or damage to property arising out of or relating to the use of this book, including any of the recommendations, instructions, or materials contained therein.

PERMISSIONS
Book purchasers may photocopy, print, or e-mail up to five copies of an NSTA book chapter for personal use only; this does not include display or promotional use. Elementary, middle, and high school teachers may reproduce forms, sample documents, and single NSTA book chapters needed for classroom or noncommercial, professional-development use only. E-book buyers may download files to multiple personal devices but are prohibited from posting the files to third-party servers or websites, or from passing files to non-buyers. For additional permission to photocopy or use material electronically from this NSTA Press book, please contact the Copyright Clearance Center (CCC) (*www.copyright.com*; 978-750-8400). Please access *www.nsta.org/permissions* for further information about NSTA's rights and permissions policies.

The *Next Generation Science Standards* ("*NGSS*") were developed by twenty-six states, in collaboration with the National Research Council, the National Science Teachers Association, and the American Association for the Advancement of Science in a process managed by Achieve Inc. Chapters 6–12 were originally published online at *www.nextgenscience.org/appendix-d-case-studies*. Reprinted with permission. The *NGSS* are copyright © 2013 Achieve Inc. All rights reserved.

Library of Congress Cataloging-in-Publication Data
NGSS for all students / [edited by] Okhee Lee, Emily Miller, and Rita Januszyk.
   pages cm
 ISBN 978-1-938946-29-5
1. Next Generation Science Standards (Education) 2. Science—Study and teaching—United States. I. Lee, Okhee, 1959- editor of compilation.

LB1585.3.N53 2015
507.1'073—dc23

2015001209

Cataloging-in-Publication Data for the e-book are also available from the Library of Congress.
e-LCCN: 2015012818

CONTENTS

ABOUT THE EDITORS — viii

CONTRIBUTORS — x

PREFACE — xi

1 *Next Generation Science Standards:* **Giving Every Student a Choice** — 1
Stephen Pruitt

2 **Science and Engineering Practices for Equity: Creating Opportunities for Diverse Students to Learn Science and Develop Foundational Capacities** — 7
Helen Quinn

3 **On Building Policy Support for the** *Next Generation Science Standards* — 21
Andrés Henríquez

4 **Charges of the *NGSS* Diversity and Equity Team** — 29
Rita Januszyk, Okhee Lee, and Emily Miller

5 **Conceptual Framework Guiding the *NGSS* Diversity and Equity** — 37
Okhee Lee, Emily Miller, and Rita Januszyk

6 **Economically Disadvantaged Students and the *Next Generation Science Standards*** — 43
Members of the NGSS Diversity and Equity Team

7	**Students From Racial and Ethnic Groups and the *Next Generation Science Standards*** *Members of the NGSS Diversity and Equity Team*	61
8	**Students With Disabilities and the *Next Generation Science Standards*** *Members of the NGSS Diversity and Equity Team*	83
9	**English Language Learners and the *Next Generation Science Standards*** *Members of the NGSS Diversity and Equity Team*	101
10	**Girls and the *Next Generation Science Standards*** *Members of the NGSS Diversity and Equity Team*	119
11	**Students in Alternative Education and the *Next Generation Science Standards*** *Members of the NGSS Diversity and Equity Team*	139
12	**Gifted and Talented Students and the *Next Generation Science Standards*** *Members of the NGSS Diversity and Equity Team*	157
13	**Using the Case Studies to Inform Unit Design** *Emily Miller, Rita Januszyk, and Okhee Lee*	171

14 Reflecting on Instruction to Promote
Equity and Alignment to the *NGSS* 179
Emily Miller and Joe Krajcik

15 Case Study Utility for Classroom Teaching
and Professional Development 193
Emily Miller, Rita Januszyk, and Okhee Lee

INDEX 203

ABOUT THE EDITORS

Okhee Lee is a professor in the Steinhardt School of Culture, Education, and Human Development at New York University. Her research areas include science education, language and culture, and teacher education. Her current research involves the scale-up of a model of a curricular and teacher professional development intervention to promote science learning and language development of English language learners. She was a member of the writing team to develop the *Next Generation Science Standards* (*NGSS*) and leader for the *NGSS* Diversity and Equity Team through Achieve Inc. She is also a member of the Steering Committee for the Understanding Language Initiative at Stanford University.

Emily Miller is a practicing teacher and a member of the *Next Generation Science Standards* (*NGSS*) Elementary and Diversity and Equity writing teams. She has taught science as an ESL/bilingual resource science specialist at a Title I school for 17 years. She has used the *NGSS* in her own classroom and improved and refined teaching to the standards with her students. She is consulting with the Wisconsin Center for Educational Research to develop teacher tools that promote sense making and language learning for English language learners in science. She authored or coauthored an *NGSS* culturally responsive engineering grant, a school garden curriculum grant, and a culturally and linguistically responsive teacher training grant for her school district. Currently, she is pursuing a PhD at the University of Wisconsin–Madison.

ABOUT THE EDITORS

Rita Januszyk is a retired fourth-grade teacher from Gower District 62 in Willowbrook, Illinois. Her responsibilities have included teaching in grades K–5 and serving as the district's science coordinator and enrichment coordinator. She received a bachelor's degree in biological science from the University of Illinois at Chicago, served as a scientific assistant at Argonne National Laboratory, and received a master's degree in elementary education from Northern Illinois University. More recently, she was a member of the *Next Generation Science Standards* (*NGSS*) writing team and member of the *NGSS* Diversity and Equity Team through Achieve Inc. She is also a writer on the middle school team for the Illinois State Board of Education Model Science Resource Project.

CONTRIBUTORS

Andrés Henríquez
Program Director
Division of Research on Learning in Formal and Informal Settings
National Science Foundation
Washington, DC

Joe Krajcik
Director
Institute for Collaborative Research in Education, Assessment, and Teaching
Environments for Science, Technology, Engineering and Mathematics
(CREATE for STEM Institute)
Professor of Science Education, College of Education
Michigan State University
East Lansing, Michigan

Stephen Pruitt
Senior Vice President
Achieve Inc.
Washington, DC

Helen Quinn
Professor Emeritus of Physics
National Accelerator Laboratory
Stanford University
Stanford, California

PREFACE

OKHEE LEE

The *Next Generation Science Standards* (NGSS Lead States 2013) are being implemented when critical changes in education are occurring throughout the nation. On one hand, student demographics across the country are changing rapidly and teachers have seen the steady increase of student diversity in the classroom, while achievement gaps in science and other key academic indicators among demographic subgroups have persisted. On the other hand, the *NGSS* and the *Common Core State Standards* (CCSS), in English language arts and mathematics are spreading. As these new standards are cognitively demanding, teachers must make instructional shifts to prepare all students to be college and career ready. Furthermore, as the standards are internationally benchmarked, the nation's students will be prepared for the global community.

The *NGSS* offer both opportunities and challenges for educators in enabling all students to meet the more rigorous and comprehensive standards set forth by the *NGSS*. The *NGSS* indicate performance expectations of students by blending science and engineering practices, crosscutting concepts, and disciplinary core ideas. Most science teachers are unaccustomed to teaching for three-dimensional learning and will be compelled to make adjustments in their instruction.

The *NGSS* have addressed issues of diversity and equity from the inception. The *NGSS* Diversity and Equity team takes the stance that the standards must be made accessible to all students, especially those who have traditionally been underserved in science classrooms, hence the title "All Standards, All Students." Through the two-year process of the *NGSS* development, the team completed four major charges: (1) bias reviews of the *NGSS*, (2) Appendix D on diversity and equity, (3) inclusion of the topic of diversity and equity across appendixes, and (4) seven case studies of diverse student groups.

Within the broader scope of the team's charges, this book focuses on the seven case studies written by the team members who are classroom teachers. The case studies are an attempt to pilot the vision presented in *A Framework for K–12 Science Education: Practices, Crosscutting Concepts, and Core Ideas* (hereafter referred to as the *Framework*; NRC 2012) and the *NGSS* with respect to diverse student groups across grade levels and science disciplines. These case studies illustrate how teachers blend the three dimensions of the *NGSS* with effective classroom strategies to ensure that the *NGSS* are accessible to all students. Furthermore, they provide practical and tangible routes toward effective science instruction with diverse student groups.

PREFACE

Each case study consists of four parts. First, it starts with a vignette of science instruction to illustrate learning opportunities through effective classroom strategies and connections to the *NGSS* and *CCSS ELA* and *CCSS Mathematics*. The vignette emphasizes what teachers *can do* to successfully engage students in meeting the *NGSS*. Second, it provides a brief summary of the research literature on effective classroom strategies for the student group highlighted in the case study. Third, it describes the context for the student group—demographics, science achievement, and educational policy. Finally, it ends with an *NGSS*-style foundation box for a user-friendly review of the *NGSS* and the *CCSS* that were taught in the vignette.

The vignettes in the seven case studies were modeled after those of *Ready, Set, Science! Putting Research to Work in K–8 Science Classrooms* (Michaels, Shouse, and Schweingruber 2008)—which is a companion to *Taking Science to School: Learning and Teaching Science in Grades K–8* (NRC 2007)—as a precursor to the *Framework*. Both sets of vignettes authenticate ideas about science education through classroom trials. However, the vignettes in this book differ in several ways: (1) they represent seven diverse demographic groups of students within the same volume; (2) they illustrate the blending of science and engineering practices, crosscutting concepts, and disciplinary core ideas; (3) they include research-based classroom strategies to improve access of diverse student groups to the NGSS; (4) they are extensive, ranging from two weeks of science instruction to an entire school year; and (5) they span K–12 grade levels and include all science disciplines.

While the book focuses on the seven case studies, we expand its scope by including seven additional chapters. The book begins with contributions by Stephen Pruitt (Chapter 1), Helen Quinn (Chapter 2), and Andrés Henríquez (Chapter 3). Then, we describe the team's charges (Chapter 4) and our conceptual framework to guide the readers in how to interpret and apply the case studies across classroom contexts (Chapter 5). The main body of the book includes the seven case studies (Chapters 6 through 12), to be followed by professional development considerations and a reflection guide for each case study. Next, we offer suggestions about how teachers can draw from case studies to inform their unit design by incorporating important shifts to support student learning (Chapter 13). Finally, Joe Krajcik, in collaboration with Emily Miller, introduces a teaching rubric that assists reflection on three-dimensional learning and focuses on equity (Chapter 14). By keeping a balance between the case studies and chapters, we maintain the integrity of the *NGSS* work on the case studies while further enhancing the team's work to make it more relevant and applicable to the broader education system.

The book makes significant contributions in several ways. First, the case studies in the book are an integral part of the development of the *NGSS*. Content standards across subject areas are written for all students, but the specific opportunities and

PREFACE

demands that are extended to diverse student groups through rigorous standards have never been similarly addressed. Second, educational research tends to address diverse groups separately but not collectively as this book does. Third, the book benefits from the combination of teacher, expert, and "teacher-as-expert" voices. Teacher-practitioners offer invaluable insights into implementation of the *NGSS* with diverse student groups, adding authenticity to the claim of utility for science educators. Finally, the book provides the context for each student group in terms of demographics, science achievement, and educational policy.

This book is intended for K–12 science educators, science supervisors, leaders of teacher professional development, education researchers, and policy makers. The primary audience of the book is classroom teachers. We encourage them to make instructional shifts in implementing the *NGSS* with diverse student groups who have historically not met district and state goals in science. In addition, this book is intended for science supervisors and professional development providers to offer support systems for classroom teachers. Furthermore, this book serves as a guide for teachers, supervisors, or professional development providers to design action plans for the *NGSS* implementation with diverse student groups. Through this publication, the case studies may reach a broad audience and initiate dialog about how to enable all students to achieve the academic rigor of the *NGSS*.

We would like to acknowledge many individuals who contributed to this book. First of all, we appreciate those individuals who contributed to the case studies:

1. *Economically Disadvantaged Students:* Rita Januszyk wrote the case study. The vignette is based on the video of the teaching of Bethany Sjoberg, Highline Public Schools, Seattle, WA. The video came from Windschitl, M., J. Thompson, and M. Braaten (2008–2013). *Tools for ambitious science teaching.* National Science Foundation, Discovery Research K–12, *http://tools4teachingscience.org*. Joseph Krajcik and Cary Sneider, both *NGSS* writing team members, collaborated on the vignette.

2. *Students From Racial and Ethnic Groups:* Emily Miller wrote the case study. She worked with Susan Cohen, a middle school science teacher at Madison Metropolitan School District and planned the curriculum with Leith Nye, Great Lakes Bioenergy Resource Center, Wisconsin.

3. *Students With Disabilities:* Betsy O'Day, *NGSS* Diversity and Equity Team member, wrote the case study.

4. *English Language Learners:* Emily Miller wrote the case study. She planned the unit with Nick Balster, University of Wisconsin–Madison, and taught the unit with her team members Stacey Hodkiewicz and Kathy Huncosky, Madison Metropolitan School District, Wisconsin.

PREFACE

5. *Girls:* Emily Miller wrote the case study. The vignette is based on the teaching of Georgia Ibaña-Gomez, School District of Cambridge, Wisconsin, and curriculum planning with Cheryl Bauer Armstrong from the Earth Partnership for Schools at the University of Wisconsin–Madison. Cary Sneider, *NGSS* writing team member, collaborated on the vignette.

6. *Students in Alternative Education:* Bernadine Okoro, a member of the *NGSS* Diversity and Equity Team, wrote the case study in collaboration with Emily Miller.

7. *Gifted and Talented Students:* Rita Januszyk wrote the case study.

In addition to Betsy O'Day and Bernadine Okoro, we also would like to acknowledge Jennifer Gutierrez and Netosh Jones, two additional members of the *NGSS* Diversity and Equity Team.

We would like to acknowledge the support of the editorial team members of NSTA Press. Claire Reinburg has guided us from the inception of the book proposal along with Wendy Rubin, Managing Editor, and Andrew Cooke, Senior Editor, who have provided valuable editorial support. We would also acknowledge Ted Willard, NGSS@NSTA Program Director, for his encouragement and vision. Finally, we would like to acknowledge Bilal Dardai, who provided excellent editorial assistance of the draft manuscript.

REFERENCES

Michaels, S., A. Shouse, and H. Schweingruber. 2008. *Ready, set, SCIENCE! Putting research to work in K–8 science classrooms*. Washington, DC: National Academies Press.

National Research Council (NRC). 2007. *Taking science to school: Learning and teaching science in grades K–8*. Washington, DC: National Academies Press.

National Research Council (NRC). 2012. *A framework for K–12 science education: Practices, crosscutting concepts, and core ideas*. Washington, DC: National Academies Press.

NGSS Lead States. 2013. *Next Generation Science Standards: For states, by states*. Washington, DC: National Academies Press. *www.nextgenscience.org/next-generation-science-standards*

Windschitl, M., J. Thompson, and M. Braaten. 2008–2013. *Tools for ambitious science teaching*. National Science Foundation, Discovery Research K–12. http://tools4teachingscience.org

CHAPTER 1

NEXT GENERATION SCIENCE STANDARDS
GIVING EVERY STUDENT A CHOICE

Stephen Pruitt

INTRODUCTION

The *Next Generation Science Standards* (*NGSS*) represent an incredible opportunity for students in science education. The *NGSS* are the result of a two-step process that was led by states and included literally thousands of educators, scientists, and other stakeholders. The first step was to develop *A Framework for K–12 Science Education: Practices, Crosscutting Concepts, and Core Ideas*, led by the National Research Council (2012) in partnership with the American Association for the Advancement of Science and the National Science Teachers Association. The scientific community would set the science standards all students should learn to become scientifically literate based on extensive research. Why would so many care enough to participate in the development of a set of science standards? What opportunities are available through science education? The answer is simple—science education offers opportunities for success in college, career, and life, but only if it is offered to every student. If every student is not given the opportunity to engage in a quality science education, then as an education system, we relegate some students to a substandard education. It is time for this trend to end. It is time for "All Standards, All Students."

A SPUTNIK MOMENT, OR SOMETHING BIGGER?

In October of 1957, the U.S.S.R. successfully launched Sputnik. The effect on science education was significant. In the span of a few years, there was a great focus on math and science education. Everyone wanted to be a scientist or engineer. The nation was focused on reaching the Moon before the end of the decade. We were able to unite around this common goal. This was great news for the United States but unfortunately not for all students in the country. Science was seen as an elite subject. Despite claiming that the United States needed to focus on science and mathematics, the reality was that only a small population had full access. Mainly middle class white males benefited from that era of science education.

CHAPTER 1

We now live in a technological society that is, at least in part, the result of the Sputnik age. Yet we still have students who do not have access to a quality science or math education. We now stand at a critical juncture. Publication after publication shows the lack of preparedness of students to enter science, technology, engineering, and mathematics (STEM) careers. Additionally, the United States continues to lag behind other industrialized countries on international assessments. These issues have been documented and discussed over and over again. The "leaky" STEM pipeline is a problem, but not having students to go into the pipeline is a bigger problem. Students of all backgrounds, ethnicities, and economic situations should have the opportunity to choose to enter these fields. This cannot happen if students are never engaged in a quality science and engineering education. To be clear, our goal should not be to make everyone a scientist. This will not and should not happen. However, significant portions of the U.S. population cannot even see themselves in STEM careers because they feel science is reserved for some kids, not all.

The *NGSS* offer an opportunity to give access to all. So, I believe the *NGSS* are not simply the 21st century's "Sputnik moment." I would argue they are bigger: The *NGSS* are the attempt to give *every* student scientific literacy and subsequent opportunity. When Dr. James Gates made the motion for the Maryland State Board of Education to adopt the *NGSS*, he gave his endorsement by saying that the *NGSS* will allow access to all students and open the world of opportunity. This is the promise of the *NGSS*—allowing access to a quality science education for all students allows access to the American dream and to global competitiveness and citizenry.

SCIENCE, THE GREAT EQUALIZER

Students in the 21st century have greater opportunities than any in history. Well, some do. As we continue to struggle with literacy and numeracy, many forget the need for scientific literacy. Why is scientific literacy so important? Is this something that scientists and science educators push because they love their discipline or is there something deeper? The ability of a student, or future adult, to think scientifically is critical in this global society. But it also provides opportunities to many underrepresented groups. Students from all backgrounds benefit by engaging in a quality science education that requires scientific thought and engagement through scientific and engineering practices or what has been historically known as *inquiry*. All students see increases in achievement when engaged in a quality science instruction, and nonmainstream and less privileged populations see greater increases in achievement than their mainstream and more privileged counterparts.

Obviously, to get along in the 21st century, students need literacy and numeracy in the traditional sense: reading, writing, and mathematics. However, understanding and engaging in scientific thought processes can help students accelerate their competencies in literacy and numeracy as well as scientific literacy. Of course, that is assuming we do not reduce science to a list of vocabulary words that are memorized for a test or placed on a

Next Generation Science Standards: Giving Every Student a Choice

wall. Science can be leveraged to meet students at their interest and in their own world. It allows them to make sense of the world around them even when experience or language is a barrier. Science is a great equalizer. It must be if we are to give every student an equal opportunity for adult success.

DEVELOPING THE *NGSS*

The process for developing the *NGSS* was not a simple one. There were many decisions made that while necessary did not make the process simpler. However, one decision was never in debate: The *NGSS* were meant for all students. To achieve this, the development had to have all students in mind from the beginning.

Twenty-six states came together to develop the standards. The states agreed to several commitments for this process, including naming a state team that would act not only as reviewers of the *NGSS* but also as advisors to the states on the *NGSS*. These teams ranged from 50 to 150 individuals, depending on the states. As part of these teams, the states were asked to have individuals who had training and worked with students with disabilities, English language learners, gifted and talented students, and other underserved populations. This was critical to make sure the *NGSS* would be able to grant access to a quality science education that is often not available to these groups.

Like the states, the *NGSS* writing team needed to have this same expertise. Forty-one individuals from very diverse backgrounds were selected through an application process to work with the states to develop the *NGSS* based on the *Framework*. It would have been much easier to have had only a few writers, but that would not have allowed for the diversity of perspectives needed to develop the *NGSS*. The writers were selected for the usual reasons of science disciplinary knowledge, grade bands, and science and engineering expertise. In addition to those areas, a group of educators were selected based on their expertise and daily engagement with underserved student populations to work together on the Diversity and Equity Team. In order to always keep a fresh and pragmatic perspective during development, the team would spend time together, in discipline groups, and in grade-band groups. From the beginning of this process, all students had to be a priority, not an afterthought. The expertise of this group was critical to bringing science to all students. Of all the decisions made by all the different players in this development, I am probably most proud of this one.

The work began with the team members working within disciplinary and grade-band groups. This was important because the first six months were spent getting a deep understanding of the vision laid out in the *Framework*. They had to have the same level of understanding as the discipline group. Of course, it should be clearly stated the Diversity and Equity Team were not there to observe, as they had expertise in the science classrooms as well. The *Framework* and ultimately the *NGSS* required such a change in science education that all writers had to have a deep understanding of the changes. At the same time, the

CHAPTER 1

team began to develop a process for how they would ensure that the *NGSS* were targeted and accessible to every student.

The team developed a bias review process of the *NGSS*. This robust process was necessary for the students who would eventually work toward the *NGSS* but also for states that would consider adoption. Several states have requirements to do bias reviews themselves, and the process developed by the team served as a guide to states and as evidence of the importance placed on access for all. However, that was not enough.

The vision for this team from the beginning was to provide guidance to the field. It was not enough to say the *NGSS* went through a bias review or to simply spout off statistics or research around science education. There needed to be a real focus that gave practical and usable information on how to engage students of all backgrounds in science. The whole point of standards, and why I choose to work in this area, is to ensure that every student has an opportunity to choose her or his own way in life. Even so, the best standards are useless if there is no guidance on how to engage all students.

In working with the Diversity and Equity Team, we decided that in addition to what would become Appendix D: "All Standards, All Students," there needed to be specifically targeted examples of engaging students from all backgrounds in science education. The appendix itself was always intended, as it was critical that we look at science education through all lenses. The charge to the team was to structure this appendix in order to give the usual demographics, science achievement trends and gaps, and research literature. In addition to the appendix, the charge included the provision that the appendix could not read as just a research project, but had to include practical examples of how to engage each of the diverse student groups.

We identified the need to discuss seven groups of students: economically disadvantaged students, students of different racial and ethnic backgrounds, students with disabilities, English language learners, girls, alternative education students, and gifted and talented students. As the work on the appendix proceeded and states reviewed the work that was put together, the states pushed for extended works around each group. The team began to discuss vignettes for each group. Quickly, the vignettes became more in-depth. As the team worked with each group, they leveraged their own expertise and that of colleagues as they moved the vignettes into the actual case studies. It took a long time to prepare the case studies and many rounds of reviews before they were finally released, but they have proven critical to making the case that science is for all students.

When the case studies were first introduced, many had questions regarding their purpose and utility. As stated earlier, they were reviewed multiple times as well as discussed in focus groups. The concerns ranged from practical to political in nature. At the end of the day, however, people began to see the utility of really focusing on access and equity beyond the demographics, achievement, and research literature. The case studies in Appendix D hold great promise. They are both a point of pride for the team and one of

Next Generation Science Standards: Giving Every Student a Choice

the most discussed components of the *NGSS*. States, districts, and schools are using the appendix and the case studies as professional development opportunities. They are being used as guides for the review of effective classroom strategies and insights into access and equity of all students in science education.

A PERSONAL NOTE

To be clear, the appendix and the case studies were done because the needs of all students were paramount in the *NGSS* development. They were not done just because it is the "in" thing to do in education. I have been asked fairly often why I got into the standards business. It is a pretty simple answer: I believe we should allow every student to have the opportunity to do whatever she or he wants to do in life. The *NGSS* are meant to prepare students for college and career, but they are also meant to ensure scientific literacy. That is to say, a student who is able to succeed under the *NGSS* will be able to choose her or his postsecondary direction. While many make great points regarding the "leaking STEM pipeline," my issue is that we do not have enough students going into the pipeline in the first place. Again, I am not saying that we should make all students scientists and engineers; rather, I am advocating that we give them the choice.

It should be unconscionable for a student, or adult for that matter, to say, "I do not like or understand science." In the 21st century, it is just as problematic as saying, "I do not like or understand math," or "I cannot read." It is a reality that one cannot get better at something by doing it less. Science is no different. By the same token, we have to recognize that engaging *all* students in science must move beyond the memorization of terms and processes and into genuine engagement. Science is not only able to meet students where they are, but it also can move them farther down the path toward achievement.

So, why do I work with standards? Because every child, regardless of circumstance, deserves the best chance at the life he or she wants. Education is the true equalizer in our society. I am very proud of the work of this team to make the *NGSS* come to life and show its connections to all students. It is my hope that all students have the opportunity to make their own life choices. The *NGSS* is the vehicle that brings science to all students. As my friend Dr. Gates said in his motion for the Maryland State Board of Education to adopt the *NGSS*, science existed before the *NGSS*, but the *NGSS* make a quality science education available to all.

Without further delay, enjoy the rest of this book, written by people who have a great passion for ensuring students have access to a quality science education. We all believe in "All Standards, All Students."

CHAPTER 2

SCIENCE AND ENGINEERING PRACTICES FOR EQUITY

CREATING OPPORTUNITIES FOR DIVERSE STUDENTS TO LEARN SCIENCE AND DEVELOP FOUNDATIONAL CAPACITIES

Helen Quinn

This chapter serves two purposes. First, it provides a context for the work presented in this book from the perspective of the chair of the committee that developed *A Framework for K–12 Science Education: Practices, Crosscutting Concepts, and Core Ideas* (hereafter referred to as the *Framework*; NRC 2012), the guiding document that underlies the *Next Generation Science Standards* (*NGSS*; NGSS Lead States 2013). Second, it goes beyond the *Framework* to discuss my personal view that science classrooms aiming to meet these new standards are an ideal place to support students in developing a set of foundational capacities that underlie much of later educational success. Supporting this development is a key element of supporting equity of opportunity to learn. All students have these capacities, but their level of development depends on experiences. Differences in this development due to differences in opportunity across the demographic spectrum are a significant source of differences in educational outcomes.

THE VISION OF THE *FRAMEWORK* AND *NGSS*: SCIENCE FOR ALL STUDENTS

New standards embody new challenges, but they also create new opportunities. The *Framework* and the *NGSS* provide an opportunity to make science learning accessible and engaging to a broader spectrum of students than have traditionally been well served in this regard. This is no accident. Both the developers of the *Framework* and the writing team that translated this document into the *NGSS* were convinced that science is a subject that every student today should have the opportunity to learn. Improving the equity of that opportunity was consciously and clearly part of the agenda for both teams.

GOALS OF SCIENCE FOR ALL STUDENTS

There are multiple reasons why every student needs to learn science. In the K–12 years, science education must serve three goals: (1) developing a science-literate population,

CHAPTER 2

(2) preparing career readiness independent of career choice, and (3) attracting some students to head for careers in science and engineering and related fields.

First, every adult today encounters issues, both in their community and in their individual life, when they must make decisions that can be—indeed, that *need* to be—informed by an understanding of science concepts and science practices. Equity demands that every student has the opportunity to develop the capabilities and understanding they will need to make life and citizenship decisions that involve interpreting scientific information and data. Whether the question is what treatment to use for their child's medical problem or where to place a new wastewater treatment plant in their community, every adult in today's world confronts many such decisions. Everyone needs an ability to apply the science practices as they interpret the available evidence in reaching a conclusion. Science does not typically provide the answer, but it can help in understanding the likely outcomes of various possible choices and thus in making a more informed decision.

Second, employability in the modern world depends more and more on having broad capabilities in 21st-century skills, such as the ability to solve unfamiliar or unexpected problems, work in teams, or respond to changing circumstances. The development of these skills can be supported in the science classroom when instruction is designed to engage students in doing science rather than just hearing about it. Science and engineering practices are aspects of desired 21st-century skills, such as the ability to make decisions based on evidence, communicate clearly, and use computers and other technology and instruments effectively. Science and engineering practices also engage students in working with others as a team, require them to listen to the ideas and arguments of others and respond flexibly to new information when revising and improving their models, explanations, or designs. These are skills and abilities that employers point to as critical for today's demanding workplace. Equity of opportunity to develop these skills and abilities in science classrooms is an important element of equity in the workforce, today and in the future.

Finally, we want some students to become fascinated by science or engineering and to go on to careers in these fields. Equity further demands that all students have the opportunity to learn science in a way that offers them that possibility, should they choose it. To broaden the diversity of the population of those careers, we must start by offering high-quality and engaging science learning opportunities to all students, regardless of race, home language, gender, disability, or parent income and education level. Science learning for all is an important component of an education designed to support students in developing the skills, understandings, and knowledge to live healthy and productive lives in today's society, and also of any effort to broaden the diversity in science and engineering careers. Thus, all students need rich opportunities to learn science, starting in the earliest grades and continuing through high school.

Science and Engineering Practices for Equity: Creating Opportunities for Diverse Students to Learn Science and Develop Foundational Capacities

THE VISION OF SCIENCE FOR ALL STUDENTS IN THE *FRAMEWORK* AND *NGSS*

The committee that developed the *Framework*, of which I was the chair, embraced the vision of science for all students as a foundation of its work. This vision was informed by research on learning in general, and science learning in particular, including research on issues of equity. Many choices made by the *Framework* committee—for example, to stress the need for science learning in K–12 as well as later and to emphasize engagement in both science and engineering practices as an integral part of science learning—were informed by knowledge of what is most effective in supporting science learning that is deep and lasting and, at the same time, inviting to a broad range of students.

The development of the *NGSS* likewise included a strong emphasis on the fact that these standards are for every student, not just a privileged few. The case studies in this volume are a part of the work of the *NGSS* development. The *NGSS* Diversity and Equity Team created these case studies as prototypes to support and illustrate the vision of the *Framework*—namely, that the approach to science education embodied in these standards provides opportunity to learn for all students—by implementing a set of trial *NGSS*-oriented units in diverse classrooms, each with a focus on a particular group of students.

The *Framework* and *NGSS* begin from a vision of science for all students and promote an approach to achieving that vision based on research on science learning (The *Framework* and the NRC study *Taking Science to School* (2007) both provide multiple references to the research that underlies this claim). What that research tells us is that students learn science best when they deeply engage in the practices of science and engineering and apply these practices to progressively (over multiple years) develop a coherent understanding of a set of disciplinary core ideas and connections between those ideas across science disciplines, labeled as crosscutting concepts in the *Framework* and *NGSS*.

We envision a science classroom where students are actively engaged with natural or engineered phenomena, where they apply their science knowledge to develop models of the system in question and use those models as they seek to explain the observed phenomena or develop designs that apply their science knowledge to solve engineering problems. In such a classroom, students talk with one another and with their teacher about science and engineering ideas, listen to the ideas of their peers as well as to the questions and explanations suggested by their teacher, and read and write about science as an avenue to further develop their understanding. Clearly a first step for all students to learn science is to offer all students the opportunity to experience such classrooms and a coherent science curriculum, starting during their elementary years and extending throughout their high school years. Thus, monitoring and ensuring that such opportunities are equitably provided to all students is a first step to achieving the "All Standards, All Students" goal of the *NGSS*.

CHAPTER 2

The stress on science and engineering practices and crosscutting concepts plays another role: It stresses what is common across all of science. The common emphasis and common language for these practices and crosscutting concepts across the science curriculum help students see science as a connected body of knowledge and develop a more coherent understanding of the nature and uses of that knowledge. Students who recognize this coherence and applicability find science learning more meaningful and more attractive. Supporting such a perspective is found to be particularly important for low-income and minority students and also for girls to become interested in science. It also supports deeper learning and persistence, even for the students who traditionally do well in science. Studies of entering college students who intend to major in science or engineering find that a measure of where the students place their view of science in the continuum—between a connected and applicable body of knowledge at one end and a collection of disconnected and not very useful facts at the other end—is a better predictor of who will complete those majors than their prior science and mathematics grades.

FALSE DICHOTOMIES RELATED TO EQUITY ISSUES

I want to pause here to make sure that I address two possible misconceptions of the nature of the science education that the *Framework* and *NGSS* demand. These misconceptions arise from false dichotomies and distract and detract from understanding the messages of the *Framework* and *NGSS*. These false dichotomies have particular significance for equity issues in science classrooms.

The first false dichotomy is between serious learning and activities that are interesting to students. When we say that science education should be engaging, that does not mean that it is all just about having fun, playing around, or messing around with stuff. Students engaged in the science and engineering practices indeed do some "hands-on" work, but even more importantly they must have their minds on the job. They may be engaged and fascinated by the phenomena introduced and they may enjoy the challenges of developing explanations of what is observed, but they are also working hard and learning new ideas. All too often at-risk students are offered only didactic teaching because this false dichotomy leads educators, whether the teacher or the principal, to believe that "these students" do not have time to "play around" but must be pushed forward by direct instruction and repeated practice using worksheets. However, learning research suggests the opposite, namely that at-risk students may have the most to gain from learning through engagement in science and engineering practices.

The second false dichotomy is between students developing their own explanations of phenomena and students being taught well-established ideas of science. The stress on students developing their own models for systems or explanations of phenomena does not mean that they are expected to rediscover the great ideas of science. Engaging in the practices does not replace learning science ideas; rather, it serves as the vehicle for

Science and Engineering Practices for Equity: Creating Opportunities for Diverse Students to Learn Science and Develop Foundational Capacities

understanding them more deeply. One cannot develop a model or argue from evidence about nothing; these practices must be used in the context of some core idea. Indeed, core ideas of science are effectively understood when they are learned in the context of trying to explain some natural or engineered phenomenon where they apply. Students incorporate the science ideas into a model for the system being examined, use the model to explain the observed phenomenon, or design a next experiment to test a proposed explanation. The phenomenon or design problem must be carefully chosen so as to confront the contradictions inherent in typical naïve conceptions of the world, to raise questions for which students see that they need new answers, and to help students see why the scientific understanding provides a better explanation than their prior conceptions. This work helps students to process the science ideas and to make the conceptual shifts needed to truly understand and incorporate the science ideas into their way of looking at the world. All students need the opportunity to engage in this work of conceptual change, and at-risk students again may have the most to gain by doing it simply because they are less likely to have had out-of-school exposure to these concepts or activities.

The science learning called for in the *Framework* and *NGSS* is demanding and rigorous. The case studies discussed in this volume demonstrate this, and at the same time demonstrate how teachers can support classrooms of diverse students in meeting those demands. These case studies, and other more rigorous research studies, demonstrate that a wide range of students, when exposed to science learning through engagement in science and engineering practices and use of crosscutting concepts, can make very significant gains in their science capability, while at the same time advancing language, literacy, and mathematics capabilities in significant ways.

FOUNDATIONAL CAPACITY DEVELOPMENT THROUGH SCIENCE AND ENGINEERING PRACTICES

Science education in the early grades can also serve as key to educational opportunity more broadly. This is because science education of the type suggested by the *Framework* and *NGSS*, offered equitably to all children starting in kindergarten, can play an important role in fostering the development of a set of capabilities that I call "foundational," because they provide a foundation for academic success. My ideas here are a synthesis, based on the work of many leading researchers, particularly those who study early childhood development. They also owe much to conversations with my daughter, who is the Director of a Spanish-immersion preschool in San Francisco (Centro Las Olas), which follows a philosophy based on the Reggio Emilia model. We identify four foundational areas of capacity development that are critical for academic success: (1) language, (2) analysis and reasoning, (3) representation and symbolization, and last, but by no means least, (4) social and emotional capacity, including self-regulation, persistence, motivation, and mindset.

CHAPTER 2

Many studies have focused on the role of developing one or another of these capacities in supporting academic success. My analysis of these studies has convinced me that all of them are critically important. For example, basic literacy skills depend on language being recognized as well as on decoding the symbols that represent the words of that language. The words themselves represent not only concrete objects and actions but also more abstract concepts and ideas. Thus, literacy builds on both the development of language and of representation and symbolization capacities. Higher-level language arts standards also require students to analyze text and reason about the author's intent in choosing particular words or style. Similarly, mathematics capabilities build on recognition of symbols and other ways of representing quantities as well as analysis and reasoning about relationships between them. All of these skills are more readily developed by students who possess certain important social and emotional skills of self-regulation, persistence, and belief in self-efficacy in learning these skills. However, these foundational capacities are generally not the targets of standards-based accountability testing. Nevertheless, aspects of learning that standards-based accountability does test and monitor cannot be stably developed without attention to the foundational capacities.

All children have these foundational capacities, but the degree to which they become developed, and the timeline of that development, varies widely. Development depends critically on the opportunities and experiences that the child is exposed to (NRC 2007). In our society, many of these opportunities strongly correlate with parent income and education. Whenever issues of equity arise, gaps in achievement between students from different socioeconomic groups or of different ethnic or racial backgrounds loom as a challenge. These gaps are well documented and persistent. A substantial part of these gaps can be traced back to differences in early childhood experiences, particularly to differences in development of these foundational capacities at the point when children enter school (Burkam and Lee 2002). Hence, an important element of achieving equity is that schooling—especially in the earliest grades and preschool but continuing throughout the K–12 years—must attend to supporting broad development of these capacities as well as to achievement of particular standards for learning. Higher-grade academic outcomes, even on the explicit learning measures related to standards, are supported by broad capacity development. Furthermore, in the longer view, employment and life path are dependent on such development. One of the unintended outcomes of a stress on standards and on scores on performance measures narrowly linked to those standards has been to divert instruction, particularly in schools serving disadvantaged populations, to modes that may achieve short-term gains on such measures in early grades but that fail to support the development of the very capacities that lead to success on more demanding tasks in later grades (Almon and Miller 2011).

The science classroom that attends to developing these foundational capacities serves diverse students well, not just to further their science learning but to support their broader academic progress. Fortunately, the classroom designed to achieve outcomes such as those called for by the *NGSS* is inherently well positioned to play this role. This is one of the

Science and Engineering Practices for Equity: Creating Opportunities for Diverse Students to Learn Science and Develop Foundational Capacities

messages of the case studies described in this book. Accepting the premise that participating in a science practice-rich classroom is the way students best learn science, let us explore the aspects of this classroom that promote development of the foundational capacities, and hence support better outcomes across the academic spectrum. Young children are naturally curious, and a science classroom where students engage in science and engineering practices around interesting phenomena can capture that curiosity and build learning experiences that channel and extend that curiosity, thereby supporting both foundational capacity development and science learning. Below, I will address each of the four foundational areas of capacity development.

LANGUAGE

First, the science practice–rich classroom provides abundant language development opportunities, not just for those who are designated as English language learners, but for every student. In such a science classroom, students are using their experience of real-world phenomena to learn new language. Their participation in classroom discourse supports learning to express ideas, thoughts, and questions effectively; to having them contribute to the shared development of ideas; and to making decisions about what to do next to resolve contradictions or to revise a model or explanation. Students become engaged in the work of science learning through discourse-laden science practices (Lee, Quinn, and Valdés 2013). Students' desire to share their ideas and contribute their questions motivates not just their science learning but their language development as well. As they develop their ability to read science-related textual materials and to write about their science thinking with support and practice in the science classroom, as well as through the tasks assigned for them to complete outside of classroom time, they are also developing language and literacy skills.

It has become evident that wide differences in language development prior to entering school (Fernald, Marchman, and Weisleder 2013) are one of the key factors in differences in educational outcomes for different groups of students. But even the most articulate five-year-old must continue to develop language ability across years of schooling, so language development is a key part of education for all students. Attention to this aspect of learning is even more important for students who enter school with limited language development. Those who enter school with well-developed language other than the language of the school (whether it is not English or nonstandard English) also need particular attention to help them build on the base of their first language(s) to develop a second (or third) that will serve them well in the school, and later in their likely work setting.

Modern theories of language development stress that participating in conversations, having multiple opportunities to speak and to be attended to, and listening to a variety of other speakers, particularly peers, are critical elements in language development, especially for children (Valdés 2004). One key to what motivates such participation is the interest and engagement of the language learner. Well-designed science lessons provide activities and

CHAPTER 2

discussions that prompt that engagement, and at the same time introduce the learners to words and concepts that stretch their vocabulary and their tools for expressing their ideas. To make science thinking clear, the speaker must communicate about specific aspects of the system under study and attend to fine details of what is being observed and discussed. Students engaging in science discourse must refine their language to communicate the ideas they have formulated. The attempt to do so drives them to clarify their thinking as well as to stretch their language capacity. Their desire to contribute their ideas can help overcome the barrier that typically inhibits productive use of newly learned language. Essentially all of the science and engineering practices require student discourse to be a central element of classroom activity and, properly managed by the teacher, such discourse includes all students and pushes every student to refine and extend language abilities.

Not only spoken language but science reading and writing require and support language development while also being crucial elements of science learning. The science teacher must support students to be effective users of science text, with its multimodal presentation and its many subject-specific peculiarities in grammatical structure, vocabulary, and density of detail. Strategies taught for extracting the main point of a paragraph in language arts may not always be useful for reading science, where much of understanding requires attention to precise details and distinctions. Students need to learn not only new words but also recognize that words have multiple meanings and to distinguish the technical meaning of a word (such as *force* in physics or *family* in biology) from its everyday meaning. New strategies to coordinate multiple sources of information and multiple modes of presenting information (through diagrams, graphs, tables or charts and equations) must be learned. Students must likewise learn to present their own ideas and models through oral and written discourse. All of this and more are implied in the science practice "obtain, evaluate and communicate information." Engaging in this practice supports students to develop literacy as well as language for science. For more information on literacy in science see *Literacy for Science: Exploring the Intersection of the* Next Generation Science Standards *and* Common Core for ELA Standards, *a Workshop Summary* (*www.nap.edu/openbook.php?record_id=18803*).

ANALYSIS AND REASONING

Moving on to the capacity for analysis and reasoning, this too is demanded, and hence advanced through its use, as students engage in science and engineering practices. One learns to reason and analyze by being asked to do it, and engagement in the full set of science and engineering practices requires this experience. When students are asked to refine questions, to develop models and model-based explanations, to argue from evidence, to plan investigations, to analyze data, to use mathematics and computational thinking, to define constraints on a design solution and develop a solution that fits within those constraints, and to evaluate the sources and reliability of information, they are being asked to reason for themselves and to analyze their ideas and those of others. They must reason and

Science and Engineering Practices for Equity: Creating Opportunities for Diverse Students to Learn Science and Develop Foundational Capacities

argue from evidence to refine and improve their science models or engineering designs, or to critique those of others.

Teachers prompt students with questions that invoke the reasoning behind their ideas, or ask them to argue whether the evidence supports or contradicts an idea or a proffered explanation of a phenomenon. This opportunity must be given in a climate that supports a focus not on who is right and who is wrong, but on what each can add to the discussion that helps the group arrive at a shared understanding. The role of the teacher in setting this climate and in asking questions chosen to prompt careful analysis and reasoning is critical to ensure that this opportunity is offered to all students, not just to those quickest to express their ideas. This process supports science learning that is better integrated into the student's conceptual viewpoint, and hence both more readily used in new situations and more likely retained, than science learning acquired by being told what is so and asked to accept it on the authority of the teacher or the text. Such experiences also support development of the capacity and disposition to analyze and reason both inside and outside the science classroom.

REPRESENTATION AND SYMBOLIZATION

Representation is a key part of science, and learning to use and interpret a wide variety of representational diagrams and simulations is key to science learning (Latour 1990). This is one reason students are asked to develop and refine models that represent the system that they are studying and to use their models to support their explanations of the phenomena observed. Science models employ both diagrams and symbolic codes to represent both seen and unseen elements of the system, such as arrows for forces, or bar graphs that represent energy stored in or flowing into or out of the system. Even visible features may be encoded symbolically; for example, mountains and hills represented by contour lines on a map. In the early grades student experience in interpreting a story told by a sequence of diagrams or cartoons is an important precursor to reading a written story. The same skill is developed as students model the steps in an observed process by a series of sketches or seek to make diagrammatic instructions for building the object that they have designed. As students are asked to develop, discuss, and refine models for the system under study or the one they wish to build, they develop their capacity to use and interpret a wide variety of such symbolic representations. They discuss and critique their own and other students' models in terms of the value in developing an explanation or solving an engineering problem.

As students learn to interpret, understand, and appreciate the value of abstract diagrammatic models, they can better follow the models or computer-based simulations presented by the teacher to help explain particular science concepts. To represent phenomena, models must capture and represent changes over time, typically through multiple versions of the system diagram, each representing a specific instant in the time progression of the

CHAPTER 2

phenomena, or a dynamic progression of such diagrams in a computer-based simulation. Science models may include aspects in addition to the system diagrams, such as a map of where a phenomenon or species was observed, a graph or equation that represents a measured relationship between some variables in the system, a key that provides a way to decode symbols used in the diagrams, and labels that emphasize and name certain key elements of the diagram. Student ability to develop and to interpret such models requires practice, starting with simpler and more concrete cases in the earliest grades, and working up to more abstract representations in the later grades. Practice in developing and interpreting models develops both language and analysis capacities as well as representational and symbolization capabilities.

The ability to read science texts requires facility with the multiple modes of representation of ideas used and must be supported by science teachers, for example, leading a discussion of how the information represented in the diagrams, graphs, or equations can be interpreted and coordinated with the information given in the verbal text to obtain the full picture of the ideas that are being presented. The conventions of representation are particular to each field of science, and teachers must support their students so they may recognize and use them as an integral part of their science learning. All this is very science specific, but the ability to abstract and represent and to reason using such representations that is developed in science learning activities carries over to learning in other content areas.

EMOTIONAL AND SOCIAL CAPACITY

The final foundational capacity is emotional and social capacity. Development of this capacity is required for students to perform well in the type of classrooms that the *NGSS* will create, and overall has an important impact on academic outcomes. Thus it is important that teachers recognize the emotional and social demands and conditions in their classrooms, and support students in developing the behaviors and attitudes to become successful learners. Similar demands exist in other subject areas from the *Common Core State Standards, ELA* and *Mathematics*. This work is a challenge for all teachers, but it is an integral piece of what must be attended to to support effective learning for all students. Students recognize that the culture of the classroom is different from their home or playground culture, but that difference must be managed and explained so all students feel at home in the classroom, welcomed as full members, and invited to share their ideas and participate fully in the activities that are being offered to promote learning.

Orchestrating a classroom where students engage in science practices requires the teacher to attend to the social and emotional dimension in that classroom, and to support students to develop their social skills and their emotional capacities and attitudes toward learning across the years of school. In such a classroom students work independently in small groups much of the time, discussing ideas amongst themselves and carrying out activities and investigations chosen by the teacher to support the intended learning or

Science and Engineering Practices for Equity: Creating Opportunities for Diverse Students to Learn Science and Develop Foundational Capacities

designed by the student to pursue a question that has arisen. The teacher moves from group to group, questioning, prompting thought, and managing interactions among students, while sometimes asking for the attention of the full class to a single speaker. The teacher ensures that discourse is inclusive and that contributions are respected and acknowledged, no matter who expresses them or with what level of language ability they have been expressed. The students come to share a common goal to understand something, to solve the puzzles presented by the phenomenon, and to apply science ideas in their models, arguments, and investigations. Sometimes a new understanding is reached for more students by discussing and eventually eliminating a wrong idea about the system in question than by accepting the right idea because the teacher says it is correct. If they have not seen why their initial idea is wrong, they will quickly revert to thinking that way when asked to apply their knowledge, even if they can state the "right" answer on a test of memorized knowledge.

Students need support to understand that it takes self-regulated behavior to participate, and that the social norms of behavior in the classroom are there to support all students to learn and to enjoy that learning experience. Teachers need to establish these norms and motivate students to follow them. Self-regulation is just one aspect of what students must practice and thus learn in such a classroom.

Effective participation in this type of learning reflects a growth mindset about learning, a belief that engagement and effort can impact learning outcomes. This mindset is fostered by a teacher attitude that accepts mistakes as part of the process, values questions as much as answers, and values ideas not for whether they are right or wrong but for the way they advance the thinking of the group. When students learn that initial failure of an engineering design can be overcome by analyzing the causes of that failure and revising the aspects of the design that led to it, they have learned an important lesson about persistence. The teacher who believes and communicates these ways of viewing learning can support students to develop a growth mindset that fosters taking on challenging problems and persisting in the face of initial failure. This growth mindset promotes more effective learning for a broad range of students than a mindset of fixed abilities, where students think of themselves and their peers as intrinsically strong or weak learners in a particular subject area (Dweck 2007).

CONCLUSION

A critical aspect of achieving effective science learning for diverse learners is that all students are supported not just to participate but to become engaged and motivated to learn science. This participation is encouraged by linking students' science-learner identities with their cultural and social identities and also with their perceptions about science and scientists. Students make decisions whether they should attend to science learning based on multiple and subtle cues about for whom it is relevant. The diversity of problems to which science is applied can and should reflect the diversity of student cultural backgrounds and issues relevant to their communities. Diverse historical science personalities can be introduced and their contributions discussed. Furthermore, teachers recognize contributions couched in informal or imperfect English, or coming from different cultural perspectives, as an expected and welcome part of the dialogue in the science classroom. The teachers whose classrooms are glimpsed in the case studies described in this volume provide examples of such teaching. As teachers implement engagement in science and engineering practices in their classrooms, they would do well to consider these examples of how to structure the culture of their classrooms and how to make teaching moves to achieve successful science learning for all students.

REFERENCES

Almon, J., and E. Miller. 2011. *The crisis in early education: A research-based case for more play and less pressure.* New York: Alliance for Childhood. *www.allianceforchildhood.org/sites/allianceforchildhood.org/files/file/crisis_in_early_ed.pdf*

Burkam, D. T., and V. E. Lee. 2002. *Inequality at the starting gate: Social background differences in achievement as children begin school.* Washington, DC: Economic Policy Institute: Research and Ideas for Shared Prosperity.

Dweck, C. S. 2007. *Mindset: The new psychology of success.* New York: Random House.

Fernald, A., V. A. Marchman, and A. Weisleder. 2013. SES differences in language processing skill and vocabulary are evident at 18 months. *Developmental Science* 16 (2): 234–248.

Gopnik, A., A. N. Meltzoff, and P. K. Kuhl. 1999. *How babies think: The science of childhood.* New York: William Morrow and Company.

Latour, B. 1990. Drawing things together. In *Representation in scientific practice*, ed. M. L. Lynch and S. Woolgar, 19–68. Cambridge, MA: MIT Press.

Lee, O., H. Quinn, and G. Valdés. 2013. Science and language for English language learners in relation to *Next Generation Science Standards* and with implications for *Common Core State Standards for English language arts and mathematics. Educational Researcher* 42 (4): 223–233.

Metz, K. E. 1995. Reassessment of developmental constraints on children's science instruction. *Review of Educational Research* 65 (2): 93–127.

National Research Council (NRC). 2007. *Taking science to school.* Washington, DC: National Academies Press.

National Research Council (NRC). 2012. *A framework for K–12 science education: Practices, crosscutting concepts, and core ideas.* Washington, DC: National Academies Press.

National Research Council (NRC). 2013. *Monitoring progress toward successful K–12 STEM education.* Washington, DC: National Academies Press.

NGSS Lead States. 2013. *Next Generation Science Standards: For states, by states.* Washington, DC: National Academies Press. *www.nextgenscience.org/next-generation-science-standards*

Valdés, G. 2004. Between support and marginalization: The development of academic language in linguistic minority children. *International Journal of the Bilingualism and Bilingual Education* 7 (2 & 3): 102–132.

Science and Engineering Practices for Equity: Creating Opportunities for Diverse Students to Learn Science and Develop Foundational Capacities

ACKNOWLEDGMENT

I would like to thank my daughter Elizabeth (Bethica) Quinn for sharing with me her ideas about foundational capacity development and allowing me to use them in this chapter. Furthermore, her dedication to bilingual development for her own children and for the children in her school has deeply influenced my interest in and thinking about language development issues, and thus contributed to those aspects of this chapter.

CHAPTER 3

ON BUILDING POLICY SUPPORT FOR THE *NEXT GENERATION SCIENCE STANDARDS*

Andrés Henríquez

In this chapter I discuss what I observed as a foundation officer at Carnegie Corporation of New York throughout the development of the *Common Core State Standards* (*CCSS*) and the *Next Generation Science Standards* (*NGSS*). In particular, I will focus on how the topic of diversity and equity was woven into the *CCSS* and the *NGSS*.

THE MOTHER OF ALL POLICY WINDOWS: *COMMON CORE STATE STANDARDS*

"Everything seems to be in place," I recall saying on a conference call with colleagues from the Carnegie Corporation of New York around 2007. We were discussing a new and exciting partnership that was taking place with the National Governors Association Center for Best Practices (NGAC) and the Council of Chief State School Officers (CCSSO). NGAC and CCSSO were undertaking an effort to work with governors and chief state school officers to build the case and establish a collaborative network of states that would have common standards in English language arts (ELA) and mathematics to be shared among the states.

At the time, NGAC and CCSSO were building their vision for the development of a common set of standards based on two ongoing models. The first was Achieve Inc.'s America Diploma Project (ADP). Achieve had been working with a network of states, including governors' offices, state education officials, and businesses to align "school standards, graduation requirements and assessment and accountability systems with the demands of college and careers" (*www.achieve.org/adp-network*). ADP included standardized ELA and mathematics curriculum, which were shared among 15 states, steadily increasing to include 35 states covering 85% of students in public schools.

The second model that influenced the process informing the *CCSS* was being implemented at about the same time as the first: the New England Common Assessment Program (NECAP). A number of states, including Maine, New Hampshire, Rhode Island, and Vermont had partnered for several years, collaborating on common assessments in reading, writing, mathematics, and science at various grade levels.

CHAPTER 3

These two models made a compelling case for policy makers and offered them freedom to experiment with a process for embarking on a shared set of college- and career-ready standards and assessments for grades K–12—previously considered a "third rail" issue. The "third rail" status stemmed from politicians' fear that national standards would eventually lead to a national curriculum and create a federally run education system. There had been previous standards efforts by earlier federal administrations. For example, President Clinton introduced the Goals 2000: Educate America Act, which called for a set of national standards as well as a voluntary set of assessments. The Act received substantial opposition from the GOP and was eventually abandoned.

This new standards movement arose from a "perfect storm" set of events that had been quietly brewing over several decades and that abruptly and massively coalesced at a crucial moment in 2008. At this time, the country was dealing with two wars, a financial crisis, and a newly elected president who promised major reforms in education and science when he took office in January 2009. The financial crisis was putting states in a desperate situation for funding teacher salaries, pensions, and other educational costs. One way states could overcome these financial straits was to save resources by collaborating with one another. This situation of mutual goal setting and collaboration spurred conversations around the development of a common set of standards. The idea quickly became an enticing notion and with the financial stimulus to states and schools from the American Recovery and Reinvestment Act of 2009 (ARRA), the policy window for a shared set of standards was thrown wide open.

One of the more convincing reasons for building a set of shared standards was the focus on equity. If the United States was going to continue as a global leader, its citizenry would need to be highly educated. This requirement extended to the majority of citizens, not just a subgroup. It could no longer be acceptable for children in Massachusetts to vastly outperform children in Mississippi. Likewise, it could no longer be acceptable to base predictions of student future achievement and opportunity solely on zip code. Parents and policy makers both agreed that there was a grave disparity in the quality of education children were receiving in U.S. schools. The push for more rigorous, wide-reaching standards could be a powerful mechanism toward mitigating disparities, if resources and opportunities are provided for all students to meet these standards.

Building on this new sense of flexibility, there was excitement and hope in the education and government sectors in early 2009. Federal funding of Race to the Top, as well as various philanthropic support, encouraged most states to adopt the *CCSS, ELA* and *Mathematics*. This remarkable accomplishment of the *CCSS* opened the door for a similar effort toward more rigorous standards in science.

On Building Policy Support for the *Next Generation Science Standards*

LAUNCHING THE *NEXT GENERATION SCIENCE STANDARDS*

As the platform for ELA and mathematics standards was established, policy makers' attention turned to science. One of the first efforts the Carnegie Corporation decided to fund was the National Academies of Sciences to develop a new science framework—a compilation of the new research and advancements in science as well as in science learning. The Board on Science Education at the National Academies of Sciences, chaired by Helen Quinn, produced *A Framework for K–12 Science Education: Practices, Crosscutting Concepts, and Core Ideas* (NRC 2012).

In addition, there was great interest in leveraging the substantial groundwork that had been laid with policy makers throughout the *CCSS*. The conversation at Carnegie Corporation, one of the many philanthropic groups that contributed to the *CCSS* and *NGSS*, was cautiously optimistic about elevating the importance of science that allied groups had been working on in a concerted way for the past 20 years. Our question was: Given the policy window and the "readiness" of policy makers around standards, could the public accept yet an additional set of standards—this time in science? Was this a moment that could be seized?

Carnegie Corporation had just established a commission with the Institute for Advanced Study (Carnegie/IAS Commission) to establish a new path for K–12 education in mathematics and science. Unlike the domain of mathematics, which had remained relatively unchanged over the century, the field of science and the way science was taught in schools had undergone enormous changes since the launch of Sputnik. Education research and our collective knowledge of how people learn have supported dramatic shifts in pedagogy, and science education was in desperate need of attention (Bransford, Brown, and Cocking 2000). In the past, science was thought of as a discipline for the "brightest" students, but access to science for every student could substantially change the way our citizenry engaged with science.

The first public report released by the Carnegie/IAS Commission was entitled *The Opportunity Equation: Transforming Mathematics and Science Education for Citizenship and the Global Economy*. In the report, the Commission found that far too few students in the United States received high-quality instruction in mathematics and science, and subsequently, the nation was falling behind many countries in these areas. In many states, science standards emphasized large amounts of content without context, lacked a coherent structure, and had little focus on scientific reasoning, experimentation, and inquiry. The report outlined the need for new mathematics and science standards and urged states and school systems to support new standards. The report also recommended an effort to prepare a teaching force that could teach toward higher mathematics and science standards (Carnegie Corporation of New York and Institute for Advanced Study 2009).

There were a number of differences between the *CCSS* and the *NGSS*. For example, the *CCSS* was written by a small group of experts, and there was consultation with external advisors before the *CCSS* were rolled out. The *NGSS* had a large number of writers

CHAPTER 3

representing various sectors of science and science education, and engaged in a much more interactive process with states and teacher teams before its launch. Other differences are discussed below.

"GOING AT IT ALONE": THE *NGSS* WERE NOT TIED TO FEDERAL SUPPORT FOR STANDARDS.

The catalyst for the widespread adoption of the *CCSS* by 44 states was Race to the Top, funded though the ARRA, providing over $100 billon for emergency education funding to states. Additionally, the U.S. Department of Education funded two consortia of states and organizations, which received substantial federal funding to develop common assessment systems that accompany the *CCSS*. The *NGSS* had generous support from Carnegie Corporation and other philanthropic groups, but had no federal support for development, dissemination, or adoption. The *NGSS* were developed by a coalition of 26 lead states that participated in multiple rounds of writing sessions. It was thought that states that worked on the effort were more likely to adopt the standards.

STATE-LEVEL WORK LEADING TO EQUITY AND BUY-IN BY STATES

The adoption process of the *NGSS* was a state-led effort with 26 lead state teams that helped to shape the work of the *NGSS*. For this reason, the subtitle of the *NGSS—For States, By States*—accurately describes the development of the *NGSS*. Teams of teachers from the states met to discuss the practices, core ideas, and crosscutting concepts, and made certain that there was high fidelity to the *Framework*. In addition, public drafts were made available on two separate occasions, and the 26 lead state teams responded to the public comments. This process of bringing in voices committed to equity at multiple levels—including states, the public, and business partners—was instrumental in achieving the vision of equity that had been put forth from the onset.

PARTNERSHIPS

Building partnerships was essential to the *Framework* and the *NGSS* effort. There were already a number of players with a different sense of what science standards were. To ensure that a number of organizations worked and collaborated with one another, there were a number of key players funded by Carnegie Corporation to ensure that the *NGSS* were launched and considered by state leaders, and had broad participation among critical stakeholders in the science policy community. It is important to note that the timing for these grants was not linear; instead, they were funded in parallel with the *NGSS* effort.

- Achieve Inc.: Besides developing the science standards, Carnegie Corporation and other philanthropic groups also supported Achieve to partner with NGAC and CCSSO to develop *Benchmarking for Success: Ensuring U.S. Students Receive a World-Class Education*, which provided states a roadmap for benchmarking their K–12 system against top-performing nations and demonstrated what it would take for U.S. students to have a world-class education (NGAC, CCSSO, and Achieve 2008).

- Council of Chief State School Officers (CCSSO): CCSSO was funded by Carnegie Corporation to provide technical assistance to states for the adoption of the *CCSS* and to ensure that future standards, including science standards, were favorable to business as well as state education agencies.
- Council of State Science Supervisors (CSSS): As part of a Carnegie Corporation grant to NSTA, there was a subcontract to CS^3 ("CS cubed") to put together a high-end advisory board to strengthen the infrastructure of the organization. CS^3 members are involved in their respective states and are proactive in improving science education. CS^3 members have been helpful partners as the *NGSS* are introduced to various state legislatures.
- Fordham Institute: Fordham undertook an analysis of state-by-state science standards. Fordham Institute's publication *The State of State Science Standards 2012* further demonstrated to the governors and the business community the need for a strong set of science standards for states (Lerner et al. 2012).
- Hunt Institute: Hunt Institute was funded to help state education policy decision makers learn about, adopt, and take the necessary steps to implement the science standards that Achieve Inc. prepared.
- National Science Teachers Association (NSTA): A project was funded to build Classroom Opportunities Multiply with Practices and Application of Science Standards (COMPASS), an online collaborative learning environment designed to provide both face-to-face and online professional development to support the active engagement of the nation's science educators.
- National Association of State Boards of Education (NASBE): NASBE worked in partnership with Achieve and other partners to educate state members in preparation for the adoption of the *NGSS*.
- National Research Council (NRC): NRC coordinated and wrote *A Framework for K–12 Science Education* (2012) and the more recent *Developing Assessments for the Next Generation Science Standards* (2014).

ALL STANDARDS, ALL STUDENTS

All of the various groups working on the science standards were cognizant of the student demographic shifts taking place in classrooms across the country. One important shift was marked by a growing number of English language learners in the schools. Also, issues of race and ethnicity, along with gender, students with disabilities, and students from economically disadvantaged backgrounds had received attention in the past, but were often overlooked in science education. The broadening of participation in science was seen as critically important to our nation's future in science and science education.

CHAPTER 3

Although the call for science standards was indeed an opportunity for the nation to address equity, such standards could make a greater impact with more depth because it was addressing the equity issue head-on. A commitment to equity, no matter how well meaning, is much harder for states and districts to put forth without explicit guidance. The case studies highlighted in this book are the guidance requested by states to illustrate enacting equitable practices. They are illustrative examples of how teachers can enable diverse student groups to embrace science.

PERSONAL THOUGHTS

In the 12 years that I spent at Carnegie Corporation, much of our work focused on excellence and equity. I was personally gratified that the *NGSS* were addressing the issues of equity. The alarmingly high number of children failing in our schools, particularly students of color, was distressing. Early on at Carnegie Corporation we launched an adolescent reading and writing initiative after becoming aware of the huge discrepancy of literacy rates among states (McCombs et al. 2005). The gap was much larger than anyone had imagined. What was particularly disturbing was the difficulty students were having reading complex texts outside of English language arts classes. Reading and writing for science is a critical skill for high school students who want access to college and careers. Students who are English language learners have even more barriers accessing high-level content. This gap in discipline-specific literacy comprehension, based largely on demographic status, has barred entry to high-level science and mathematics courses at the high school.

This focus of the work at Carnegie Corporation was to highlight the issue of adolescent literacy to policy makers. (Middle and high school teachers had known about this gap for years.) This was also a very personal issue for me: As an eager college student, I desperately wanted to major in biology. However, I was poorly prepared for science by my inner-city urban school and I struggled. This was both frustrating and embarrassing for me, and it took years for me to learn to read complex technical texts.

I wanted to ensure students like me—students of color, bilingual, and from economically disadvantaged backgrounds—were not forced to repeat my experience in K–12 and in college. As the *Framework* points out: "Science and engineering are growing in their societal importance, yet access to a high-quality education in science and engineering remains determined in large part by an individual's socioeconomic class, racial or ethnic group, gender, language background, disability designation, or national origin" (NRC 2012, p. 280).

Of course, much work remains to be done; the most important of which is the building of science assessments and further work in the area of teacher development. These efforts—and their timing—are key to the successful implementation of the *NGSS* and the associated improvement of science learning and success by U.S. students.

On Building Policy Support for the *Next Generation Science Standards*

The funding of the *NGSS* was one significant step toward promoting excellence and equity for all students. While the increased rigor of science education through the new standards is challenging to students, teachers, and schools, all students will be better prepared for succeeding in science in college and careers. The *NGSS* and the case studies in Appendix D, put in motion by the *CCSS* and state policy and public initiatives, set the stage for transforming teaching and learning to address the crisis not only in science education, but in disparity, by a bold and persistent call for equity.

REFERENCES

Bransford, J. D., A. L. Brown, and R. R. Cocking, eds. 2000. *How people learn: Brain, mind, experience, and school.* Washington, DC: National Academies Press.

Carnegie Corporation of New York and Institute for Advanced Study. 2009. *The opportunity equation: Transforming mathematics and science education for citizenship and the global economy.* New York: Carnegie Corporation of New York.

Lerner, S. L., U. Goodenough, J. Lynch, M. Schwartz, and R. Schwartz. 2012. *The state of state science standards.* Washington, DC: Thomas Fordham Institute.

McCombs, J. S, S. N. Kirby, H. Barney, H. Darilek, and S. Magee. 2005. *Achieving state and national literacy goals: A long uphill road: A report to Carnegie Corporation of New York.* Santa Monica, CA: RAND Corporation.

National Governors Association Center for Best Practices (NGAC), Council of Chief State School Officers (CCSSO), and Achieve Inc. 2008. *Benchmarking for success: Ensuring U.S. students receive a world-class education.* Washington, DC: National Governors Association.

National Research Council (NRC). 2012. *A framework for K–12 science education: Practices, crosscutting concepts, and core ideas.* Washington, DC: National Academies Press.

National Research Council (NRC). 2014. *Developing assessments for the* Next Generation Science Standards. Washington, DC: National Academies Press.

CHAPTER 4

CHARGES OF THE *NGSS* DIVERSITY AND EQUITY TEAM

Rita Januszyk, Okhee Lee, and Emily Miller

When the *NGSS* Diversity and Equity Team presents data about science achievement gaps to groups across the country, the reactions are mixed.

- "As educators adjust to the rigor and demands of the *NGSS*, all students will perform better. So won't the gaps persist?"
- "My level 3 ELLs will not understand these new standards."
- "It's so depressing."
- "We have lower expectations for students in poverty."
- "There is an inequality of resources contributing to inequality of access."
- "Start the process of removing barriers."
- "We have a unique opportunity to get it right!"
- "All students can learn and it begins with me."

Beginning with *A Framework for K–12 Science Education: Practices, Crosscutting Concepts, and Core Ideas* (NRC 2012) and continuing through the *Next Generation Science Standards* (NGSS Lead States 2013), the science learning of diverse student groups is part of our national discourse. Uniquely, the *NGSS* from its inception ensured that concerns of diverse student groups were addressed. Of the 41 *NGSS* writing team members, a smaller group of those individuals representing diverse student groups was brought together to form the Diversity and Equity Team. The team was given the task of ensuring that diverse student groups' interests were incorporated into the *NGSS* planning, writing, rewriting, revising, adding, subtracting, and visualizing.

With feedback from key state partners and two rounds of public comments that comprised the comprehensive *NGSS* review process, the *NGSS* writers went through an iterative process of revising and refining the standards based on that feedback. Meanwhile, the team made adjustments accordingly and completed four significant tasks: (1) Appendix D "All Standards, All Students": Making the *Next Generation Science Standards* Accessible to All Students; (2) bias reviews of the standards; (3) inclusion of the diversity and equity topic in the appendixes; and (4) seven case studies (see Figure 4.1, p. 30).

CHAPTER 4

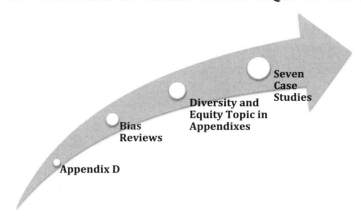

FIGURE 4.1.

CHARGES OF THE *NGSS* DIVERSITY AND EQUITY TEAM

This chapter summarizes the four tasks of the team. By understanding the far-reaching charges to address diversity and equity across many related topics and products, readers will appreciate how the interests of diverse student groups were taken into account throughout the *NGSS* writing process. In particular, the team members wrote the seven case studies so readers can find utility in the case studies for their teaching practices.

APPENDIX D

Appendix D is designed as a reference tool for classroom teachers, school and district administrators, and state policy makers. The appendix lays out current issues around equity of science education for diverse student groups.

THE *NGSS* OFFER BOTH OPPORTUNITIES AND CHALLENGES FOR ALL STUDENTS.

Consideration of learning opportunities and challenges are particularly important for student groups that have traditionally been underserved in science classrooms. In the *NGSS*, the rigor of disciplinary core ideas is intertwined with science and engineering practices (the "doing" part of science) and crosscutting concepts (bridging ideas that encompass multiple science disciplines). This *three-dimensional learning* defined by the *Framework* and embodied by the *NGSS* promises to be an opportunity for students to explain natural phenomena, solve design problems, and develop lasting scientific knowledge about how scientists and engineers really do science and engineering in their careers. Additionally, the *NGSS* provide connections to the *Common Core State Standards* (*CCSS*) and the overlap between science and engineering practices with English language arts (ELA) and mathematics practices allow for reinforcement across subject areas. For example, the science and engineering practice of Engaging in Argument From Evidence coincides with constructing viable arguments in ELA and math.

EFFECTIVE CLASSROOM STRATEGIES HELP TEACHERS WITH STRATEGIC DECISIONS.

Appendix D presents effective classroom strategies as identified by the current research literature. The strategies are an illustrative selection, rather than an exhaustive list. The common themes of these strategies are summarized as (1) valuing students' background knowledge in home or community contexts, (2) connecting background knowledge (cultural or linguistic) to school learning, and (3) building off of resources to support student learning.

The entry points for providing equity to science education are multiple (see Figure 4.2). Advocating for programs that foster parent involvement and connecting to homes requires funding and fortitude. Many good examples of school and home partnerships can be found. Informal school opportunities, such as creating alliances with museums, nature preserves, and universities (to name a few) provide connections, relevance, and support. But it often comes down to the availability and allocation of resources.

FIGURE 4.2.

EQUITABLE LEARNING OPPORTUNITIES

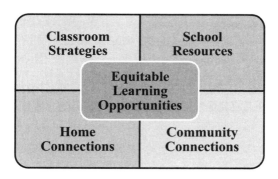

THE CONTEXT OF SCIENCE LEARNING MATTERS.

Appendix D identifies the context of science learning in terms of student demographics, science achievement, and education policy. The student population in our nation is becoming more diverse racially and linguistically and more students are living in poverty. Meanwhile science achievement gaps among demographic groups persist, according to the National Assessment of Educational Progress (National Center for Education Statistics 2011). Unanswered questions remain. If current trends in student demographics continue, what will happen to science achievement? What will it take to close the gaps? In education

CHAPTER 4

policy, No Child Left Behind (NCLB 2001) defines accountability for reading and math, but not for science. The fact that science is not part of annual yearly progress has major implications.

BIAS REVIEWS OF THE NGSS

In the development of the *NGSS*, the team reviewed the standards for bias. From the early stages of standards development, writers were alerted to potential bias possibilities. For example, a clarification statement in a performance expectation that included an example of baking cookies to describe the molecular nature of a gas might bias a student who never experienced baking cookies in her or his family life. Examples in clarification statements were carefully selected to avoid such bias.

The team reviewed the standards for consistency of language to ensure that (1) the language used in performance expectations was consistent as the concept built across grade bands and (2) language used in science and engineering practices and crosscutting concepts was consistent within a grade band and across grade bands. The team also reviewed the standards for clarity of language, while being mindful that as performance expectations translated into assessments, clarity was important for understanding the intent of the *NGSS* writers and not confusing to diverse student groups. Therefore, suggestions were made to use accessible language.

Finally, the bias review included recommendations directly impacting diverse student groups. Where appropriate, the team suggested real-world problems or reference to local contexts that are motivating to students in poverty, and low-cost materials for students in urban or rural schools. The team also suggested references to familiar objects or tools for English language learners (ELLs) as they can more effectively demonstrate their knowledge with such references.

DIVERSITY AND EQUITY TOPIC IN APPENDIXES

NGSS Volume 2: Appendixes (NGSS Lead States 2013) is the companion volume, consisting of Appendixes A through M. Many of the appendixes address diversity and equity connections and stay with the theme that the *NGSS* are for all students. The inclusion of the diversity and equity topic in the appendixes is described below.

- Appendix F: Science and Engineering Practices states that science classroom discourse is vital to three-dimensional learning and presents both language demands and language learning opportunities. This is particularly true for ELLs, speakers of nonstandard English, and students with Individualized Education Program (IEP) who have language processing difficulties: "When supported appropriately, these students are capable of learning science through their emerging language and of comprehending and carrying out sophisticated language functions (e.g., arguing from evidence, providing explanations, developing models using less-than-perfect English)" (NGSS Lead States 2013).

- Appendix G: Crosscutting Concepts emphasizes that connections through crosscutting concepts promote a more complete understanding of science for students who may have traditionally been relegated to basic level classes.
- Appendix H: Understanding the Scientific Enterprise: The Nature of Science emphasizes contributions to science and engineering by individuals and teams of men and women from diverse cultures.
- Appendix I: Engineering Design has a section called Engineering Design in Relation to Student Diversity. The NGSS elevate engineering as integral to science—a marked shift from previous standards. Engineering design is at the same level as scientific inquiry in the NGSS. Engineering permeates modern life and is important for career and college readiness. Participating in engineering problems or designing solutions creates relevance and motivation to students who otherwise might not be engaged.
- Appendix J: Science, Technology, Society, and the Environment highlights that the home/community connection to science is important for academic success of diverse groups of students who have "funds of knowledge" from their home and community contexts that can provide a bridge to science learning in the classroom (Gonzalez, Moll, and Amanti 2005).

SEVEN CASE STUDIES

The student groups for the case studies included the four federally designated accountability groups: (1) economically disadvantaged students, (2) students from major racial and ethnic groups, (3) students with disabilities, and (4) students with limited English proficiency (the federal term) (NCLB 2001). The team broadened the scope of the work to include (5) girls, (6) students in alternative education, and (7) gifted and talented students (see Figure 4.3, p. 34).

CHAPTER 4

FIGURE 4.3.
SEVEN CASE STUDIES

Economically Disadvantaged: Grade 9 Physical Science	• Developing Conceptual Models to Explain Chemical Processes
Major Racial and Ethnic Groups: Grade 8 Life Science	• Constructing Explanations to Compare the Cycle of Matter and the Flow of Energy Through Local Ecosystems
Students With Disabilities: Grade 6 Space Science	• Using Models of Space Systems to Describe Patterns
English Language Learners: Grade 2 Earth Science	• Developing and Using Models to Represent Earth's Surface Systems
Girls: Grade 3 Engineering	• Defining Problems With Multiple Solutions Within an Ecosystem
Alternative Education: Grade 10 & 11 Physical Science	• Constructing Explanations About Energy in Chemical Processes
Gifted and Talented: Grade 4 Life Science	• Constructing Arguments About the Interaction of Structure and Function in Plants and Animals

The seven case studies are from real classrooms with real teachers and students in grades 2–11. The case studies include life science, Earth and space science, physical science, and engineering. Through the vignette (story) in each case study, we enter the classroom and watch the students engage in science and engineering under the skillful guidance of a well-prepared teacher. The units or lessons include student work samples, student-to-student discourse, and student-to-teacher discourse. In the vignette you see the learning progression of the students developing conceptual understandings from their naïve or incomplete ideas to more complete ideas about the science concepts. You also see three-dimensional learning and effective strategies made explicit and annotated.

While the vignette is an authentic classroom approximation, events have been collapsed or summarized with some units ranging from a few weeks to an entire year. And the vignette is intended to illustrate students moving toward meeting standards, but is not intended to prescribe science instruction toward meeting the *NGSS*. The vignette gives teachers and administrators a vehicle for discussion and planning toward more robust curricular directions around the *NGSS*, keeping diverse students' needs as an important part of the implementation decision making. In the vignette we single out a particular diverse group for illustration, but the classrooms in the vignettes might be like yours: gifted and talented students, students from racial and ethnic groups, ELLs, students with disabilities, and others, all in the mix. More details about the case studies are discussed in Chapter 5: Conceptual Framework Guiding the *NGSS* Diversity and Equity.

CONCLUSION

Looking at the quotes at the beginning of this chapter, we hopefully will be moving from "We have lower expectations for minority students in poverty" and "It's so depressing" to "We have a unique opportunity to get it right" and "All students can learn and it begins with me." State and local organizations formulating policy and acting as agents of change can turn the pessimism into optimism.

The vision of the *Framework* and the *NGSS* to include the needs of diverse student groups to ensure access and achievement is commendable. The challenges and opportunities for diverse student groups meeting the *NGSS* are illustrated in the case studies as students are engaged in meaning making, science discourse, and explanation of natural phenomena and solution of design problems, which will lead them to the goal of career and college readiness. Appendix D and the case studies can inform discussions around addressing diversity and equity in science education.

REFERENCES

Elementary and Secondary Education Act of 1965, Pub. L. No. 89–10, 79 Stat. 27.

Gonzalez, N., L. C. Moll, and C. Amanti. 2005. *Funds of knowledge: Theorizing practices in households, communities, and classrooms.* Mahwah, NJ: Lawrence Erlbaum Associates.

National Center for Education Statistics (NCES). 2011. *The nation's report card: Science 2009.* Washington, DC: U.S. Department of Education.

NGSS Lead States. 2013. *Next Generation Science Standards: For states, by states. Volume 2: Appendixes.* Washington, DC: National Academies Press.

No Child Left Behind Act (NCLB) Act of 2001, Pub. L. No. 107–110, and115, Stat. 1425. 2002. *www2.ed.gov/policy/elsec/leg/esea02/pg1.html#sec101*

CHAPTER 5

CONCEPTUAL FRAMEWORK GUIDING THE *NGSS* DIVERSITY AND EQUITY

OKHEE LEE, EMILY MILLER, AND RITA JANUSZYK

Traditionally, subject areas such as English language arts, mathematics, and science tend to be taught independently from other subjects, especially in the middle and high school grades. In the same way, student demographic groups tend to be addressed separately (e.g., research on race or ethnicity, research on English language learners, research on students with disabilities, or research on gender). In contrast, the *NGSS* connect science disciplines and, furthermore, connect science with English language arts and mathematics. Likewise, the *NGSS* Diversity and Equity Team's work brings diverse student groups together to reflect the diversity in today's classrooms. As an outcome of this merging of science disciplines and subject areas with student diversity, the seven case studies describe connections of the *NGSS* to the *CCSS ELA* and *CCSS Mathematics* across diverse student groups.

This chapter presents our conceptual framework that guides readers to better understand the case studies presented in Chapters 6 through 12. Three issues are addressed: (1) synthesis of effective classroom strategies across the seven case studies, (2) caveats to understand the purpose of the case studies, and (3) how to approach the vignette in each case study.

This chapter is built on Chapter 4: Charges of the *NGSS* Diversity and Equity Team, with a focus on the case studies. For suggestions about how teachers can draw from case studies to inform their integrated and culturally relevant unit design, and how collaborative learning communities can use case studies as tools for making instructional shifts, readers are encouraged to read Chapter 13: Using the Case Studies to Inform Unit Design. In addition, Chapter 14 offers a teaching rubric that will inform teachers' shifts in their instructional practices recommended by Appendix D and the case studies. For a more comprehensive treatment of a theoretical and conceptual grounding for diversity and equity issues related to the *NGSS*, readers are encouraged to read Appendix D: "All Standards, All Students": Making the *Next Generation Science Standards* Accessible to All Students (NGSS Lead States 2013).

SYNTHESIS OF EFFECTIVE CLASSROOM STRATEGIES ACROSS SEVEN CASE STUDIES

The collection of seven case studies on seven demographic student groups is noteworthy. Whereas research literature on each student group tends to exist independently from the others, our work departs from this norm by addressing seven student groups together. Furthermore, whereas research literature on effective classroom strategies typically focuses on each demographic student group in isolation from the others, our work attempts a synthesis of the literature across the seven student groups.

The research literature across diverse student groups highlights that when provided with equitable learning opportunities, all students are capable of engaging in science practices and constructing meaning in both science classrooms and informal settings (NRC 2007, 2012). In our synthesis of the research literature, we identify teachers' beliefs and practices that are essential for equitable learning opportunities. Teachers should set high expectations for all of their students, and this is particularly important for the *NGSS* that are academically rigorous. Also, teachers should design science instruction around students' prior knowledge based on social, cultural, and linguistic backgrounds. In addition, teachers should be sensitive to a broad spectrum of learners by understanding patterns of a specific student group (however broadly or narrowly defined) and variations among individuals within the group.

Beyond fundamental teachers' beliefs and practices that apply to all students, we looked for the literature on each demographic student group to identify classroom strategies that are particularly effective for the group. We did not conduct a comprehensive and exhaustive review of literature; instead, we consulted with literature review articles, handbook chapters, and prominent publications by leading scholars in the field.

Effective classroom strategies across diverse student groups seem to fall into four domains: student engagement, classroom support strategies, school support systems, and home and community connections (see Table 5.1, which first appeared in Miller and Januszyk 2014). In recognition of each area of research literature, we used the terminology as it is used in the research literature. While some strategies are unique to a particular student group (e.g., home language use with English language learners, accommodations or modifications for students with disabilities), other strategies apply across student groups broadly (e.g., multiple modes of representation). Readers will find more detailed descriptions of the strategies in the case studies in Chapters 6–12.

CAVEATS TO UNDERSTAND THE PURPOSE OF CASE STUDIES

The seven case studies illustrate science teaching and learning of diverse student groups as they engage in the *NGSS*. The purpose of the case studies is to illustrate an example or prototype for implementation of the *NGSS* with diverse student groups. Each case study focuses on a particular demographic student group against the backdrop of other student groups in a regular classroom setting. Collectively, the seven case studies portray the *NGSS* implementation with the wide range of student diversity in science classrooms across the nation. As the *NGSS* are implemented in the coming years, rich and powerful examples of classroom practices that actualize the vision of "all standards, all students" will emerge.

TABLE 5.1.

SAMPLES OF EFFECTIVE STRATEGIES ACROSS DIVERSE STUDENT GROUPS

Demographic Group	Student Engagement	Classroom Support Strategies	School Support Systems	Home and Community Connections
Economically Disadvantaged Students	Students' sense of place	Project-based learning	School resources and funding	Students' funds of knowledge
Racial and Ethnic Groups	Multimodal experiences	Multiple representations; culturally relevant pedagogy	Role models and mentors	Community involvement; culturally relevant pedagogy
Students With Disabilities	Accommodations and modifications	Differentiated instruction; Universal Design for Learning; response to intervention	Accommodations and modifications	Family outreach
English Language Learners	Discourse practices	Language and literacy support	Home language support	Home culture connections
Girls	Relevance; real-world application	Curricular focus	School structure	Relevance; real-world application
Students in Alternative Education	Safe learning environment	Individualized academic support	After-school opportunities; career and technology opportunities	Family outreach
Gifted and Talented Students	Strategic grouping; self-direction opportunities	Fast pacing; challenge level	School identification programs	Family outreach

Several caveats are offered to understand the purpose of the case studies. First, for the purpose of illustration only, each vignette is focused on a limited number of performance expectations. In addition, student understanding builds over time and some topics or ideas require extended revisiting through the course of a school year or over multiple years.

Each case study highlights one identified group (e.g., economically disadvantaged students, students with disabilities). In reality, however, students do not fit solely into one demographic group, but belong to multiple categories of diversity (e.g., students with disabilities who are also racial and ethnic minorities from economically disadvantaged backgrounds). Thus, for a particular student or student group, multiple case studies could apply.

As there is wide variability among students within each demographic group, "essentializing" or stereotyping on the basis of a group label must be avoided. For example, English language learners (ELLs) form a heterogeneous group with differences in ethnic background, proficiency level in home language and English, gender, prior schooling in home country, immigration history, socioeconomic status, parents' education level, and so on. Teachers should attend to individual variations within a broad categorization of a specific demographic group.

CHAPTER 5

There is the question of whether good teaching for specific student groups is "just good teaching" for all students. Ladson-Billings (1995) argues that although good teaching is generic, "it is much more than that" (p. 159). She questions "why so little of it seems to be occurring in the classrooms populated by African American students" (p. 159) and other marginalized student groups. She also challenges "those who suggest that good teaching cannot be made available to all children" (p. 163) and advocates for culturally relevant teaching that addresses the needs of specific student groups.

Finally, there is a question of whether teachers are equally effective with different student groups. Loeb, Soland, and Fox (forthcoming) found the following results involving ELLs and non-ELLs in reading and mathematics: On average, teachers who were effective with non-ELLs were also effective with ELLs. However, teachers who spoke the home language of ELLs or possessed bilingual certification tended to produce relatively greater gains for ELLs than for non-ELLs. Consistent with these results, the research literature indicates that while teaching effectiveness tends to apply across different student groups, some teachers are more effective with some student groups than others.

HOW TO APPROACH THE VIGNETTE IN EACH CASE STUDY

Each case study presents an extensive vignette of science teaching and learning. As described above, vignettes illustrate the *NGSS* implementation by blending three-dimensional learning with effective classroom strategies for diverse student groups. The seven vignettes were written by the authors of this book and other members of the *NGSS* Diversity and Equity Team in collaboration with many individuals (see acknowledgments in Preface).

Each vignette focuses on one demographic student group (with caveats described above). The title of each vignette highlights the three-dimensional learning of the *NGSS*. For example, the title for the vignette related to ELLs, "Developing and Using Models to Represent Earth's Surface Systems," blends the science practice of Developing and Using Models, the disciplinary core idea about the Earth's surface, and the crosscutting concept of Systems. This approach applies to all of the remaining vignettes.

After a brief introduction about the purpose and setting, the vignette makes "connections" to a specific student group by highlighting members of this student group as they engaged in the *NGSS*. Across the seven vignettes, science instruction ranged from one unit of instruction for a few weeks in richer depiction to an entire school year to show a breadth of science topics and learning progression over time. Each vignette includes annotations of science and engineering practices (practice), crosscutting concepts (CC), disciplinary core ideas (DCI), and effective classroom strategies. The vignette highlights the blending of three-dimensional learning with effective classroom strategies. An example from the vignette related to ELLs is on the top of the next page.

Conceptual Framework Guiding the *NGSS* Diversity and Equity

After sharing the parent interviews and hearing Mrs. Xiong's presentation, the class was convinced that soil was different in different places, but they wanted to be sure that this was true for soil from different places in their neighborhood, too. Ms. H. tried to center her science investigations in culturally relevant contexts, in this case their neighborhood. (This "place-based" strategy established connections between school science and the students' community and lives.)

Ms. H. encouraged students to gather physical evidence for their claim that "soil was different in different places." They decided that the best way to support their claim was to observe soil taken from different places near the school (practice: Planning and Carrying Out Investigations). They used a topographical map and an aerial photo map of the neighborhood to determine soil sites that seemed different: a hill, the marsh, and the school yard. They noticed that the sites had different trees—deciduous trees, no trees, and coniferous trees—and they also had different elevations (DCI: K-2-ESS2.B Earth's Systems). It was at these sites that the students collected and investigated the soil, forming the basis for comparisons based on evidence and the soil profile diagrams each group constructed.

The following week, Ms. H. helped her students think in terms of patterns when exploring similarities and differences in the soil in the neighborhood (CC: Patterns).

In the *NGSS* Connections section, we analyze which performance expectations, disciplinary core ideas, science and engineering practices, and crosscutting concepts are addressed in the vignette. It is important to note that each vignette addresses multiple performance expectations across science disciplines (i.e., bundling of performance expectations). For example, the vignette related to ELLs bundles the following performance expectations:

2-ESS2-1 Earth's Systems: Processes That Shape the Earth
Compare multiple solutions designed to slow or prevent wind or water from changing the shape of the land.

2-ESS2-2 Earth's Systems: Processes That Shape the Earth
Develop a model to represent the shapes and kinds of land and bodies of water in an area.

2-PS1-1 Matter and Its Interactions
Plan and conduct an investigation to describe and classify different kinds of materials by their observable properties.

K-2-ETS1-1 Engineering Design
Ask questions, make observations, and gather information about a situation people want to change to define a simple problem that can be solved through the development of a new or improved object or tool.

CHAPTER 5

In the *CCSS* Connections section, we analyze how the vignette relates to *CCSS ELA* and *CCSS Mathematics*. Although many connections could be made, we highlight only those standards that have precise and meaningful connections.

CONCLUSION

As the primary purpose of this book is to support the *NGSS* implementation for diverse student groups in science classrooms, this chapter is intended to guide readers to better understand the case studies presented in Chapters 6 through 12. The chapter lays the conceptual grounding for effective classroom strategies identified for each demographic student group. It also highlights caveats to understand the purpose of the case studies. Finally, it guides the readers through how to read vignettes in case studies to fully benefit from each of the seven case studies.

The case studies can inform instruction for teachers based on their interest in science topics, student groups, and grade levels. Across the seven case studies, teachers will find familiarity in vignettes that resonate with their instructional needs or experiences. Even more, teachers may be motivated or inspired by the vignettes and apply them to further improve science instruction. As most teachers have taught diverse demographic groups of students, we recommend teachers convene with colleagues across science disciplines, subject areas, and grade levels. As teachers share ideas and collaboratively plan instruction, they may challenge one another to try new strategies for diverse learners in the science classroom. We hope readers enjoy reading and learning from the case studies.

REFERENCES

Ladson-Billings, G. 1995. But that's just good teaching! The case for culturally relevant pedagogy. *Theory Into Practice* 34 (3): 159–165.

Loeb, S., J. Soland, and L. Fox. Forthcoming. Is a good teacher a good teacher for all? Comparing value-added of teachers with their English learners and non-English learners. *Educational Evaluation and Policy Analysis.*

Miller, E., and R. Januszyk. 2014. The *NGSS* case studies: All standards, all students. *Science and Children* 51 (5): 10–13.

National Research Council (NRC). 2007. *Taking science to school: Learning and teaching science in grades K–8.* Washington, DC: National Academies Press.

National Research Council (NRC). 2012. *A framework for K–12 science education: Practices, crosscutting concepts, and core ideas.* Washington, DC: National Academies Press.

NGSS Lead States. 2013. *Next Generation Science Standards: For states, by states.* Washington, DC: National Academies Press. www.nextgenscience.org/next-generation-science-standards

CHAPTER 6

ECONOMICALLY DISADVANTAGED STUDENTS AND THE *NEXT GENERATION SCIENCE STANDARDS*

MEMBERS OF THE *NGSS* DIVERSITY AND EQUITY TEAM

ABSTRACT

The economically disadvantaged student population in the United States is growing. More than 20% of the children in the country currently live in poverty, with the greatest concentrations in cities. Despite the NAEP results for this demographic group showing steady increase in science achievement, the achievement gap between poverty and nonpoverty remains unchanged. The *Next Generation Science Standards* (*NGSS*) are more rigorous than the standards that came before. Thus, teachers of economically disadvantaged students will have to narrow the achievement gap while also meeting the higher expectations presented by the new standards. Based on the research literature, effective teaching strategies for economically disadvantaged students include (1) connecting science education to students' sense of "place" as physical, historical, and sociocultural dimensions in their community; (2) applying students' "funds of knowledge" and cultural practices; and (3) using project-based science learning centered on authentic questions and activities that matter to students. The vignette of high school chemistry instruction with economically disadvantaged students highlights how these strategies facilitate understanding of disciplinary core ideas, science and engineering practices, and crosscutting concepts as described by the *NGSS*.

VIGNETTE: DEVELOPING CONCEPTUAL MODELS TO EXPLAIN CHEMICAL PROCESSES

While the vignette presents real classroom experiences of the *NGSS* implementation with diverse student groups, some considerations should be kept in mind. First, for the purpose of illustration only, the vignette is focused on a limited number of performance expectations. It should not be viewed as showing all instruction necessary to prepare students to fully understand these performance expectations. Neither does it indicate that the performance expectations should be taught one at a time. Second, science instruction should take

CHAPTER 6

into account that student understanding builds over time and that some topics or ideas require extended revisiting throughout the course of a year. Performance expectations will be realized by using coherent connections among disciplinary core ideas, science and engineering practices, and crosscutting concepts within the *NGSS*. Finally, the vignette is intended to illustrate specific contexts. It is not meant to imply that students fit solely into one demographic subgroup, but rather to illustrate practical strategies to engage all students in the *NGSS*.

INTRODUCTION

Lincoln High School has the following demographics: free or reduced-price lunch (63.9%), English language learners (7.9%), students of color (74.8%), and special education students (13.6%). In the vignette below, the students in Ms. S.'s ninth-grade chemistry class reflect the overall demographics of the school. The students study matter and its interactions through multiple investigations about the structure and properties of matter. They are challenged to be precise with their scientific language and adapt their conceptual models as new evidence is presented. The students learn the practices and disciplinary core ideas of the *NGSS* over 14 school days of science instruction (adapted from Windschitl, Thompson, and Braaten 2008–2013). Throughout the vignette, classroom strategies from the literature that are effective for all students—particularly economically disadvantaged students—are indicated in parentheses.

As with all of the case studies in Appendix D, this unit was used in an actual classroom setting. In addition, the teaching methodology described in the vignette was a component of a research study that collected data on its effectiveness. The original case study was videotaped in a ninth-grade classroom with outcomes that now correlate to the middle school level of the *NGSS*. This shift is to be expected as schools transition to more rigorous standards. Therefore the writers have chosen to portray this vignette as originally recorded, with the caveat that the lessons should be seen as building a foundation for high school–level coursework. As with all good instruction, it is important for the teacher in the vignette to first ascertain the level of understanding that incoming students have, and then to build toward a more advanced understanding.

ECONOMICALLY DISADVANTAGED CONNECTIONS

The students in Ms. S.'s ninth-grade chemistry class built on their prior knowledge of the particle nature of matter to further explore the behavior of atoms and molecules. The learning outcomes of the unit included the concept that matter, specifically a gas, is composed of particles called molecules that move faster or slower, depending on the temperature of the gas. In addition, the students extended their learning to incorporate a relationship between the relative speed of the particles in a system and the pressure exerted on the sides of the container.

Economically Disadvantaged Students and the *Next Generation Science Standards*

The teacher promoted student learning through real-life examples and student-constructed models. She enabled the students to develop their own conceptual models, use them in predicting relationships between the model components, and evaluate them for their explanatory power (*NGSS* practice of Developing and Using Models). As the students gained understanding of the core ideas through use of the additional *NGSS* practices of Planning and Carrying Out Investigations and Obtaining, Evaluating, and Communicating Information, they addressed the limitations presented in the different models and worked together to revise the models as new evidence came to light.

Developing an Initial Conceptual Model

Ms. S. started a unit on matter and its interactions that involved analysis of the forces between atoms and molecules, but wanted to first find out if her students had an understanding of the molecular nature of matter. She used a whole-class discussion to bring out students' prior knowledge. They reviewed phase change and molecular movement in relation to temperature. Based on this informal assessment, she learned that some of the class remembered previous experiences with phase changes that occur with water.

The teacher began by asking the class to describe what they already knew about how gases behave, and related the questions to investigations that the students had completed. "We looked at air, carbon dioxide, and water vapor," she said. "What do you know about the molecules of a gas? How do they move? What affects their movement? What is a gas?" As students volunteered, she wrote down several students' responses on chart paper:

- "Gases expand when heated."
- "As a liquid evaporates, it becomes a gas and the molecules move rapidly."
- "There is a difference in density."
- "Gas is a phase."

"Molecules are small for gas and large for solid," Canyon offered. Ms. S. asked Canyon if he had any examples of his idea and he said, "No examples." She stated, "That's a question," and wrote Canyon's words on the question side of the chart paper. She added, "Does anyone want to comment on Canyon's remark?" Lorenzo contributed that he thought molecules stay the same size and that as molecules heat up, they move faster. After listing many student responses, Ms. S. asked the driving question: "How do gases and their behavior affect matter?"

Ms. S. next presented the class with a real-world scenario, using photographs and video (practice: Obtaining, Evaluating, and Communicating Information). In the video, a railroad tank car (tanker) was washed out with steam and then all the outlet valves were closed. The video revealed the tanker dramatically imploding the next day. After watching the video twice, the students began to speculate as to why the tanker collapsed. They

CHAPTER 6

thought that the car froze, exploded, or compressed, and the steam caused the tanker to collapse inward (see Figure 6.1). An understanding of the cause-and-effect concept helped students make sense of this phenomenon (*Analyzing real-world events using project-based learning is an effective teaching strategy.*) (CC: Cause and Effect).

Rick called out, "Okay, that's crazy!" Ms. S. asked the class to write in their journals their descriptions of why the tanker was crushed. "Do you want to guess?" she asked. "I have no idea," one student replied.

The teacher encouraged the class by asking them to continue to think and work in groups. Four groups of four students were created. Each group's task was to decide on one model to explain why the tanker imploded, making sure the drawings included molecules and force arrows. Ms. S. circulated among the students and asked guiding questions such as, "What happens when water vapor turns into liquid?" She directed students to include their ideas in the models they were creating. The students were drawing and discussing their models in their groups:

"Steam inside is moving fast."

"Maybe it was cold."

"Didn't explode; it imploded," clarified a student.

"Big, but sealed. Nothing in it but air and steam in there," said another (practice: Developing and Using Models).

Lorenzo decided that there was a tornado inside. Ms. S. directed the group to review what happens when steam turns into a liquid. She reminded students of a previous balloon experiment in which they had identified a pressure difference and asked, "What would cause pressure or a pressure difference?" She also encouraged students to incorporate the observation that heating a substance adds more pressure. Circulating among the four groups, she asked students about their drawings, "Why did the tanker crush the next day? How do temperature changes affect molecules? Is there pressure against the walls? Why?"

Cristiano answered, "Pressure in air is more than inside," and his partner Jasmine offered, "The steam inside turned to liquid." Ms. S. redirected their conversation with a new question, "Why would it implode?" Jasmine answered immediately, "Heat expands molecules!" "The molecules are getting smaller," contradicted Cristiano. After thinking a moment, he said, "They *don't* do that, do they?" (*Asking authentic questions in project-based learning is an effective teaching strategy.*)

Ms. S. asked the group about the air pressure arrows at the top of the tanker, "Why only at the top of the tanker?" Cristiano ventured, "There's more air on top, not at the bottom." Al added, "Molecules combine to take up less space." Ms. S. emphasized, "When molecules combine, they make new substances." Jasmine reminded the group that temperature has to do something. Ms. S. moved over to another group that had just broken into laughter and asked what was so funny. Rick related, "I see smashed cans all the time.

I think an air-foot stomped the tanker down. And the molecules transformed into a molecule foot." Ms. S. asked, "What is this imaginary foot?" Latasia answered, "Air." Ms. S. guided the students, "Let's add that idea to the model."

As the discussions continued, several students began making connections between the steam turning to liquid overnight and the resulting changes in collisions of molecules with the walls inside and outside of the tanker. Through further questioning and reminders of previous learning that contradicted students' claims, Ms. S. pressed the students to prioritize evidence while, at the same time, allowing them to generate their own incomplete conceptual model. Ms. S. was well aware that students must be allowed to construct an understanding of phenomena by putting their ideas together. She also knew that through guided experiences and meaningful dialogue students would adapt their model and demonstrate authentic learning.

Gathering New Evidence to Evaluate and Revise Conceptual Models

The following day Ms. S. encouraged students to reflect on how their ideas had evolved from the beginning of the unit. She wondered whether changes in students' ideas would be apparent in their developing models: Air molecules slow down; water changes phase to liquid; pressure arrows show the collisions of molecules against the edge of the tanker; and when the gas molecules turn to liquid, there is less pressure on the inside causing the tanker to crumple. Reviewing the question from the day before, "What would cause pressure or a pressure difference?" the class identified two key factors: temperature and pressure (DCI: MS-PS1.A Structure and Properties of Matter). The molecules that made up the steam were also hitting the inside of the tanker, balancing the air molecules hitting the tanker on the outside (practice: Developing and Using Models).

Ms. S. asked the class a new question, "What caused the pressure inside the tanker to change?" The students did not respond at first. Then Lorenzo concluded that outside air pressure pressed on the tanker to crush it. Ms. S. asked, "Why would it do that?" This

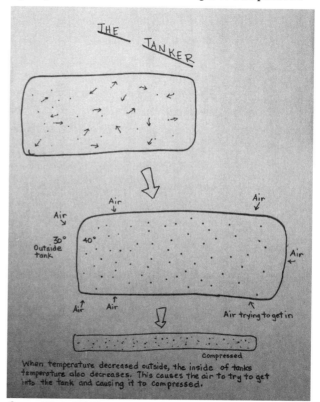

FIGURE 6.1.
STUDENTS' INITIAL MODEL

"When temperature decreased outside the inside of tanks temperature also decreases. This causes the air to try to get into the tank and causing it to compressed."

Source: Windschitl, Thompson, and Braaten 2008–2013

CHAPTER 6

> **FIGURE 6.2.**
>
> ## STUDENTS' DEVELOPING MODEL
>
> "The steam cooled down and the pressure contracted. Also the pressure decreases inside." (The bottom part of the student's writing is incomplete.)
>
>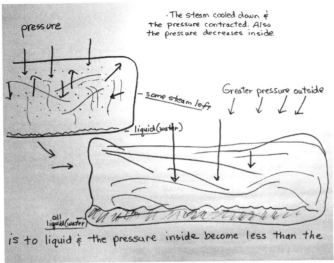
>
> Source: Windschitl, Thompson, and Braaten 2008–2013

question led Ms. S. to introduce the pop can investigation. She asked the class to make predictions, "What will happen to the pop can if water is heated inside, and the pop can is rapidly cooled?" Students called out their predictions, "It's going to do what the tanker did." "Crush!" "Implode." Jasmine asked, "Are we going to seal the container?" showing her understanding of the variables involved.

Working in their groups, the class prepared for a simulation of the crushed tanker using an aluminum pop can. The can was filled with a small amount of water, heated to boiling on a hot plate, and then submerged upside down in an ice bath using tongs. The can immediately crushed. The enthusiastic reactions from the students included:

"Oooh!"

"It's cool!"

"Awesome, it sucked it in!" *(Some comments were based on incomplete understanding.)*

The teacher asked the students to draw new models by showing the molecules of gas in the can and writing down their ideas in their science journals. *(The cultural context of the pop can was an effective use of place to connect to students' experiences in their community.)*

The following day, Ms. S. provided students with a checklist to guide their review of the can implosion investigation from the day before. The checklist included: movement of molecules (speed), phase of matter, and causes of pressure inside and outside of the can. Students were asked to write answers in their science journals. Then they discussed their ideas in groups. As she met with each group, Ms. S. pressed students to verbalize core ideas about the behavior of molecules, and left them with questions to consider. Finally, students were directed to write about their ideas so far. Ms. S. provided a scaffold for writing complete ideas by giving the class this sentence: *When _____, the can crushed more because _____.*

As their understanding grew, students refined their models and discussed changing the variables for further investigations (Figure 6.2; practice: Planning and Carrying Out Investigations). Calling the class back together, Ms. S. summarized the variables suggested by the groups: amount of water in the can, temperature of the water bath, amount of time on the hot plate, size of the can, and amount of seal when the can is flipped into the bath.

Ms. S. also reminded the students of the connection between the tanker implosion and their can implosion: The molecules of air hitting on the outside were not balanced with the molecules of steam hitting the inside.

Using Literacy, Discourse, and Argumentation to Develop a Shared Understanding

The following day the investigations continued, using students' ideas. Ms. S. asked questions as to why more steam caused more pressure. The class regrouped to perform five experiments, with each group taking one idea: amount of water, temperature of bath, time on hot plate, volume of can, and amount of seal. Each group identified three variables to test to help develop a more causal explanation. As the groups worked, the teacher questioned the students on their predictions and probed for specific answers. Lorenzo offered, "Steam vapor cools down inside the can when the can is placed in the ice bath and turns into water." "Water liquid molecules move slower than water gas molecules and the water liquid molecules take up less space because the gas condensed into water," added Jaylynn (practice: Constructing Explanations and Designing Solutions).

The group that turned the can upward in the ice water bath was surprised the can did not crush. Latasia thought there was too much space, so the can did not crush. Mia thought that with more air there was more space because of the ratio between the air and space. As shown in Mia's response, Ms. S. had identified a gap in students' understanding of pressure differences. She assigned a reading assignment on air pressure for homework (practice: Obtaining, Evaluating, and Communicating Information).

When students returned the next day, they drew a model of air pressure on people in their science journals. Alicia described her picture of pressure on Earth and pointed out that at higher elevation there was less pressure due to fewer molecules. The class reviewed the meaning of forces and how force arrows explained pressure in the model they were refining for the tanker question.

Student responses became more confident as the lessons continued. Students used a computer simulation of pressure versus. temperature and were asked to predict what would happen; the class buzzed with conversation. Next, the students improved their models. Again, students were given incomplete sentences to finish and reflect on what happened with their pop can investigations. Ms. S. reminded students to provide evidence for their explanations: "What are the molecules doing? Let's say the molecules are at a popular music concert trying to see the band. What would the molecules be doing?" Jaylynn conjectured that the quantity of molecules influenced the pressure in the can, saying, "The kids would be pushing each other to get a better view of the band. So in the can more molecules would mean less space in the can." Alicia offered, "And molecules hitting the can from the outside would not be able to push the can in." Canyon added, "When the steam cooled in the can, it meant less steam and less pressure. Because fewer molecules were hitting the inside of the can, the can collapsed." The students' responses showed they understood the concept that as the temperature decreases, the molecules move slower with fewer collisions. *(The teacher applied a cultural reference of a popular music concert, an effective strategy.)*

The students compared the results of the pop can investigations with the implosion of the tanker. As they constructed explanations, their understanding of gas behavior concepts was evident and their models were more complete. "The tanker imploded and the can got crushed because the number of air molecules hitting the outside far exceed the number of air molecules or water molecule hitting the inside." "It is the number of molecules that hit the side that causes pressure." The students concluded that under normal conditions, the tanker would not implode because the number of molecules hitting the outside would equal the number hitting the inside.

Application of Scientific Knowledge to an Engineering Problem

At the end of the two-week unit, Ms. S. challenged the teams to apply their knowledge of thermal energy and pressure to design a tanker that would not implode after cleaning. The design constraints included the use of local materials, and a feature that would ensure even poorly trained technicians would not accidentally cause the tanker to implode. Ms. S. led a discussion about how to evaluate the competing design solutions, and the class agreed upon two criteria: cost effectiveness and no implosion. (DCI: ETS1.B Developing Possible Solutions). The students were given additional pop cans to allow them to test their ideas. After about 30 minutes of small-group brainstorming, designing, and building, each group had a model to test. (CC: Structure and Function) (practice: Developing and Using Models).

Cristiano, Jasmine, and Al proposed keeping the tanker in a warm room after cleaning so that it would cool very gradually. To test their idea, they immersed it in warm water, not ice water. It imploded very slightly. Al suggested, "Let's use hot water instead of warm. Then it would cool off very slowly." The group agreed to try that.

Lorenzo's group punched a small hole at the opposite end of the can. When the can was immersed in the ice bath (with the punched hole just above the waterline), the can did not collapse at all. Lorenzo and Latasia whooped for joy! Mia reacted, "Wait! What happens to the liquid inside if there's a hole in the tanker?" "What do you mean?" asked Lorenzo. "Well, if the tanker has something like oil in it, the oil will evaporate out of the hole!" The others agreed, but liked their design anyway, and thought that the problem was not that important.

Canyon, Alicia, and Jaylynn whispered together for a long time before asking Ms. S. for materials. Jaylynn argued successfully to immerse a room-temperature pop can (not heated) in ice water. When the group tried that, the pop can did not implode. Alicia was worried, "Do you think we're cheating?" Ms. S. pointed out that it was a design worth considering and asked the group if they could think of any problems with this design. Canyon offered, "This design is great! But what if the tanker had a liquid inside that would not clean well with cold water?"

Rick's group made a sign that they said they would paint on the tank, so it would never come off. The sign said: "After cleaning, open all doors." They demonstrated how it would work by immersing the can right side up, so that cool air could flow into the tank.

Ms. S. concluded the class by pointing out that engineering problems often had many solutions, with some better than others. The next day, the groups presented their design solutions. The class discussed which of the solutions was best based on the two criteria that they had established earlier (practice: Engaging in Argument From Evidence).

NGSS CONNECTIONS

The *NGSS* vision of blending disciplinary core ideas, science and engineering practices, and crosscutting concepts is exemplified in this vignette. The learning progressions of the *NGSS* disciplinary core ideas allow teachers to assess whether students have the needed foundation for the new concepts. The teacher presented engineering practices when she introduced the tanker design engineering problem. Students were asked to apply the evidence from the pop can experiment to the real-world problem of preventing a tanker from collapsing if maintained properly. See Figure 6.3 (p. 57) for the comprehensive list of *NGSS* and *CCSS* from the vignette.

CHAPTER 6

Performance Expectations

> ### MS-PS1-4 Matter and Its Interactions
>
> Develop a model that predicts and describes changes in atomic motion, temperature, and state of a pure substance when thermal energy is added or removed.
>
> ### MS-ETS1-2 Engineering Design
>
> Evaluate competing design solutions using a systematic process to determine how well they meet the criteria and constraints of the problem.

The vignette also highlights that learning science has important implications in the real world. In the vignette, the worker who cleaned the tanks had no conceptual understanding—or at least no accurate mental model—of what would happen if he or she closed all the valves after steam-cleaning the tank. That was an expensive mistake for the company, and the worker might have lost her or his job over it. This is a lesson about the importance of science in using and maintaining equipment and illustrates the interdependence of science, technology, and engineering.

Disciplinary Core Ideas

> ### PS1.A Structure and Properties of Matter
>
> Gases and liquids are made of molecules or inert atoms that are moving about relative to each other. The changes of state that occur with variations in temperature or pressure can be described and predicted using these models of matter.
>
> ### ETS1.B Developing Possible Solutions
>
> There are systematic processes for evaluating solutions with respect to how well they meet the criteria and constraints of a problem.

Economically Disadvantaged Students and the *Next Generation Science Standards*

Science and Engineering Practices

Developing and Using Models

Use and/or develop models to predict, describe, support explanations, and/or collect data to test ideas about phenomena in natural or designed systems, including those representing inputs and outputs, and those at unobservable scales.

Planning and Carrying Out Investigations

Conduct an investigation and/or evaluate and/or revise the experimental design to produce data to serve as the basis for evidence that meet the goals of the investigation.

Constructing Explanations and Designing Solutions

Construct an explanation that includes qualitative or quantitative relationships between variables that predict(s) and/or describe(s) phenomena.

Construct an explanation using models or representations.

Engaging in Argument From Evidence

Evaluate competing design solutions based on jointly developed and agreed-upon design criteria.

Obtaining, Evaluating, and Communicating Information (by the end of grade 12)

Compare, integrate, and evaluate sources of information presented in different media or formats (e.g., visually, quantitatively) as well as in words in order to address a scientific question or solve a problem. Gather, read, and evaluate scientific and/or technical information from multiple authoritative sources, assessing the evidence and usefulness of each source.

CHAPTER 6

The students in the vignette engaged in many science and engineering practices, thereby building a comprehensive understanding of what it means to do science. The science practice of Developing and Using Models is highlighted throughout the vignette. In the course of study, the students constructed two conceptual models: the first for the tanker's implosion and the second for the implosion or lack of implosion of the pop can. The second model was more sophisticated and built on the first model, as new evidence was presented. A third model was based on the concepts from the other two and illustrated a design solution. Throughout the unit, the students were challenged to modify and revise their models as they gained an understanding of the disciplinary core ideas of the pressure and temperature variables. In addition, the students were engaged in the science practices of Planning and Carrying Out Investigations and Engaging in Argument From Evidence. In small-group and whole-group discussions, the students *constructed scientific explanations* for the tanker implosion, revised their explanations as they synthesized the tanker information, used their understanding of core ideas to construct a design solution, and supported or refuted claims. Students completed assignments by *obtaining, evaluating, and communicating information* about pressure differences and design explanations.

Crosscutting Concepts

Cause and Effect
Cause and effect relationships may be used to predict phenomena in natural or designed systems.
Structure and Function
Structures can be designed to serve particular functions by taking into account properties of different materials and how materials can be shaped and used.

The *NGSS* crosscutting concept of Cause and Effect was highlighted in the vignette as students described the effect of the forces applied on the tanker and pop can and made comparisons. The students' observations guided them to provide evidence for the causality of the tanker and pop can collapse. They made predictions about scientific phenomena based on their developing understanding of effects of molecular movement and causes for phase changes. Later the *NGSS* crosscutting concept of Structure and Function applied to the purpose of engineering a solution to prevent the implosion of a tanker.

Economically Disadvantaged Students and the *Next Generation Science Standards*

CCSS CONNECTIONS TO ENGLISH LANGUAGE ARTS AND MATHEMATICS

The *NGSS* supports an interdisciplinary approach to science learning to provide experiences across disciplines. It is for this reason that each science standard explains its connections to the *CCSS* for English language arts (ELA) and mathematics.

The high school students in the vignette grappled with core ideas in physical science while meeting the *CCSS ELA* by discussing, writing and revising explanations, and evaluating the scientific arguments presented by others.

- **RST.9-10.9** *Compare and contrast findings presented in a text to those from other sources (including their own experiments), noting when the findings support or contradict previous explanations or accounts.*
  Students had reading assignments throughout the unit: pressure and how pressure differentials are established.

- **RST.11-12.9** *Synthesize information from a range of sources (e.g., texts, experiments, simulations) into a coherent understanding of a process, phenomenon, or concept, resolving conflicting information when possible.*
  Students synthesized information from the video of the tanker collapse, their experiments, and the gas pressure vs. temperature simulation.

- **SL.9-10.2** *Integrate multiple sources of information presented in diverse media or formats (e.g., visually, quantitatively, orally) evaluating the credibility and accuracy of each source.*
  Students analyzed the simulation and compared the results of the simulation questions to their models.

- **W.9-10.7** *Conduct short as well as more sustained research projects to answer a question (including a self-generated question) or solve a problem; narrow or broaden the inquiry when appropriate; synthesize multiple sources on the subject, demonstrating understanding of the subject under investigation.*
  Investigations of the pop can questions were short research projects in which students recorded their investigation, results, and explanations.

- **WHST.9-10.1** *Write arguments focused on discipline-specific content.*
  With the help of the teacher, the students wrote arguments about their models and their learning.

The unit also addressed the grade-appropriate *CCSS Mathematics* throughout the exploration with core ideas in physical science. In the vignette the students strove to successfully combine math and science practices to present valid explanations.

- **Math Practice 2** *Reason abstractly and quantitatively.*
  In the vignette, student models reflected abstract reasoning, using a symbol system including comparisons of relative pressure.

# CHAPTER 6

- **SP** *Investigate patterns of association in bivariate data.*
  Students drew the conclusion that as one variable (temperature) increased, the other variable (pressure) increased.
- **S.IC** *Make inferences and justify conclusions from sample surveys, experiments, and observational studies.*
  Students inferred the properties of matter from their observations and experiments and justified their conclusions using the models they created.

EFFECTIVE STRATEGIES FROM RESEARCH LITERATURE

Limited resources and problems in need of fixing often frame the discourse of poverty. While these realities are part of the education landscape, they focus on deficits—what youth and their teachers and schools are lacking. Emerging literature, however, identifies resources that students bring to the science classroom and possibilities that teachers and schools can create to engage youth in science. This literature points to three effective strategies that promote science learning for economically disadvantaged students: (1) the "place" of urban and rural science education, (2) funds of knowledge and cultural practices, and (3) project-based learning to make science relevant to the students.

First, as students interact with their community, the place of science education accounts not only for the physical spaces of the community but also for the historical and sociocultural dimensions (Avery 2013; Calabrese Barton, Tan, and O'Neill 2014). The psychological, social, and physical connections to place are both sources of knowledge and critical leverage points of the emotional connection to place. Urban and rural youths' knowledge of place and their relationship with that place are powerful sources of sense-making in both formal and informal science learning environments. Alongside their capacity to navigate the physical and social spaces of their community, their scientific understandings constitute expertise that they use to make meaning of new science content, apply that content to their everyday lives, and share their knowledge of issues with community members.

Second, pedagogical practices should bridge the students' worlds with the school science world in ways that are empowering and relevant to the students. An effective teaching strategy is for teachers to validate and apply the "funds of knowledge" (González, Moll, and Amanti 2005) that students bring from their homes and communities to the science classroom. Youth have a rich store of funds of knowledge and local practices that are reflective of their cultural and linguistic backgrounds and their local communities. When teachers have understandings of students' funds of knowledge, they are able to engage pedagogical practice that fosters authentic engagement by their students. When students are provided with opportunities to leverage cultural practices in support of developing scientific knowledge and practice, they engage in higher-level scientific reasoning and participate productively in scientific inquiry.

FIGURE 6.3.

NGSS AND CCSS FROM VIGNETTE

MS-PS1 Matter and Its Interactions
MS-ETS1-2 Engineering Design

Students who demonstrate understanding can:

MS-PS1-4. Develop a model that predicts and describes changes in atomic motion, temperature, and state of a pure substance when thermal energy is added or removed.

MS-ETS1-2. Evaluate competing design solutions using a systematic process to determine how well they meet the criteria and constraints of the problem.

The performance expectations above were developed using the following elements from the NRC document *A Framework for K–12 Science Education*:

SCIENCE AND ENGINEERING PRACTICES	DISCIPLINARY CORE IDEAS	CROSSCUTTING CONCEPTS
Developing and Using Models Modeling in 6–8 builds on K–5 and progresses to developing, using, and revising models to support explanations, describe, test, and predict more abstract phenomena and design systems. • Develop a model to predict and/or describe phenomena. **Engage in Argument From Evidence** Engaging in argument from evidence in 6–8 builds from K–5 experiences and progresses to constructing a convincing argument that supports or refutes claims for either explanations or solutions about the natural and designed world. • Evaluate competing design solutions based on jointly developed and agreed-upon design criteria.	**PS1.A: Structure and Properties of Matter** • Gases and Liquids are made of molecules or inert atoms that are moving about relative to each other. • In a liquid, the molecules are constantly in contact with other; in a gas, they are widely spaced except when they happen to collide. In a solid, atoms are closely spaced and may vibrate in position but do not change relative locations. • The changes in state that occur with variations in temperature or pressure can be described and predicted using these models of matter. **ETS1.B: Developing Possible Solutions** • There are systematic processes for evaluating solutions with respect to how well they meet the criteria and constraints of a problem.	**Cause and Effect** • Cause-and-effect relationships may be used to predict phenomena in natural or designed systems. **Structure and Function** • Structures can be designed to serve particular functions by taking into account properties of different materials and how materials can be shaped and used.

CCSS Connections for English Language Arts and Mathematics

RST.9-10.9 Compare and contrast findings presented in a text to those from other sources (including their own experiments), noting when the findings support or contradict previous explanations or accounts.

RST.11-12.9 Synthesize information from a range of sources (e.g., texts, experiments, simulations) into a coherent understanding of a process, phenomenon, or concept, resolving conflicting information when possible.

SL.9-10.2 Integrate multiple sources of information presented in diverse media or formats (e.g., visually, quantitatively, orally) evaluating the credibility and accuracy of each source.

W.9-10.7 Conduct short as well as more sustained research projects to answer a question (including a self-generated question) or solve a problem; narrow or broaden the inquiry when appropriate; synthesize multiple sources on the subject, demonstrating understanding of the subject under investigation.

WHST.9-10.1 Write arguments focused on discipline-specific content.

Math Practice 2 Reason abstractly and quantitatively.

SP Investigate patterns of association in bivariate data.

S.IC Make inferences and justify conclusions from sample surveys, experiments, and observational studies.

CHAPTER 6

Finally, project-based learning is centered on authentic driving questions and activities that matter to students (Krajcik and Blumenfeld 2006). Teachers who practice project-based learning create learning environments where students socially construct knowledge based on readily available resources. It is through these connections that students who have traditionally not embraced science recognize science as relevant to their lives and future, deepen their understanding of science concepts, and develop agency in science.

CONTEXT

DEMOGRAPHICS

The American Community Survey report from the U.S. Census Bureau summarized the poverty data (2011). Overall, 21.6% of children in the United States live in poverty, the highest poverty rate since the poverty survey began in 2001. The poverty rate was the highest for African American children at 38.2% and Hispanic children at 32.3%, compared to white children at 17.0% and Asian children at 13.0%.

Students are identified by school districts as economically disadvantaged if they receive free or reduced-price lunch, or if they qualify for other public assistance (No Child Left Behind [NCLB] Act 2001). According to NCLB, poverty is measured by the number of children ages 5 through 17 who are eligible for free or reduced-price lunch under the Richard B. Russell National School Lunch Act, the number of children in families receiving assistance under the state program funded under Part A of Title IV of the Social Security Act, the number of children eligible to receive medical assistance under the Medicaid program, or a composite of such indicators. The National School Lunch Program (NSLP) is a federally subsidized program administered on local school campuses across the nation. Students are eligible for free lunch if they come from families with incomes less than 130% of the federal poverty level. Students are eligible for reduced-price lunch if their families have incomes less than 185% of the federal poverty level. For example, for the period of July 1, 2010, through June 30, 2011, for a family of four, 130% of the poverty level was $28,665, and 185% was $40,793 in most states.

According to the *Common Core of Data* report, 48% of students were eligible for free or reduced price lunch in 2010–11 compared to 47% in 2009–10 (NCES 2012a). In 2010–11, eligibility ranged among states from a low of 25% in New Hampshire to a high of 79% in the District of Columbia. The U.S. territories are also eligible for free or reduced-price lunch. Finally, a greater number of students live in poverty in the cities compared to other areas: 59.8% in cities, 39.6% in suburban areas, 51.8% in towns, and 43.9% from rural regular public elementary and secondary schools (NCES 2012a). *Low-poverty schools* are defined as public schools where 25% or fewer students are eligible for the free or reduced-price lunch (NSLP) program, *mid- to low-poverty schools* with 26% to 50% eligible students, *mid- to high poverty schools* with 51% to 75% eligible students, and *high-poverty schools* with 76% or more eligible students (NCES 2012b). The percentage of low-poverty public schools decreased from 31% in 1999 to 20% in 2010, whereas high-poverty public schools increased from 12% in 1999 to 20% in 2009. High-poverty schools are concentrated in the cities compared to the town or rural areas. In 2009–10, approximately 25% of students were in high- poverty schools (NCES 2012b).

SCIENCE ACHIEVEMENT

National Assessment of Educational Progress (NAEP) collects data on student eligibility for the National School Lunch Program (NSLP) as an indicator of family income. *The Nation's Report Card: Science 2009* looked at science performance for students in grades 4, 8, and 12 (NCES 2011). In fourth grade, 15% of students who were eligible for free lunch and 25% eligible for reduced-price lunch scored at or above proficient in science achievement, compared to 48% of students not eligible who scored at or above proficient. In 8th grade, 12% of students who were eligible for free lunch and 22% of students who were eligible for reduced-price lunch scored at or above proficient, compared to 41% not eligible who scored at or above proficient. In 12th grade, due to possible underreporting of data, students' eligibility NSLP was not included.

Economically disadvantaged students face challenges with academic success. According to a 2004 study by the RAND Corporation, socioeconomic factors such as family income, neighborhood poverty, parental education levels, and parental occupation were more significant in explaining differences in educational achievement than traditional factors such as race, ethnicity, and immigrant status (Lara-Cinisomo et al. 2004). One of the greatest challenges to schools with a high proportion of economically disadvantaged students is overcoming these challenges.

EDUCATION POLICY

Title I of the Elementary and Secondary Education Act (1965) is the largest federally funded education program. This program, authorized by Congress, provides supplemental funds to school districts to assist schools with the highest student concentrations of poverty in meeting education goals. The purpose of Title I is "to ensure that all children have a fair, equal, and significant opportunity to obtain a high-quality education and reach, at a minimum, proficiency on challenging state academic achievement standards and state academic assessments." This can be accomplished by ensuring alignment of rigorous academic standards with high-quality academic assessments, accountability systems, teacher preparation and professional development, and instructional materials. Then students, parents, teachers, and administrators can measure progress against common expectations for students' academic achievement.

CHAPTER 6

REFERENCES

Avery, L. M. 2013. Rural science education: Valuing local knowledge. *Theory Into Practice* 52 (1): 28–35.

Calabrese Barton, A., E. Tan, and T. O'Neill. 2014. Science education in urban contexts: New conceptual tools and stories of possibilities. In *Handbook of research in science education, 2nd edition*, ed. S. K. Abell and N. G. Lederman, 246–265. Mahwah, NJ: Lawrence Erlbaum Associates.

González, N., L. C. Moll, and C. Amanti. 2005. *Funds of knowledge: Theorizing practices in households, communities, and classrooms.* Mahwah, NJ: Erlbaum Associates.

Krajcik, J. S., and P. Blumenfeld. 2006. Project-based learning. In *The Cambridge handbook of the learning sciences*, ed. R. K. Sawyer, 317–334. New York: Cambridge.

Lara-Cinisomo, S., A. R. Pebley, M. E. Valana, E. Maggio, M. Berends, and S. R. Lucas. 2004. *A matter of class: Educational achievement reflects family background more than ethnicity or immigration.* RAND Corporation. *www.rand.org/publications/randreview/issues/fall2004/class.html*

National Center for Education Statistics (NCES). 2011. *The nation's report card: Science 2009.* Washington, DC: U.S. Department of Education.

National Center for Education Statistics (NCES). 2012a. *Numbers and types of public and secondary schools from the Common Core of Data: School year 2010–11.* Washington, DC: U.S. Department of Education.

National Center for Education Statistics (NCES). 2012b. *The condition of education 2012* (NCES 2012-045). Washington, DC: U.S. Department of Education.

No Child Left Behind Act (NCLB) Act of 2001, Pub. L. No. 107–110, and115, Stat. 1425. *www2.ed.gov/policy/elsec/leg/esea02/pg1.html#sec101*

U.S. Census Bureau. 2011. *Child poverty in the United States 2009 and 2010: Selected race groups and Hispanic origin.* Washington, DC: U.S. Department of Commerce.

Windschitl, M., J. Thompson, and M. Braaten. 2008–2013. *Tools for ambitious science teaching.* National Science Foundation, Discovery Research K–12. *http://tools4teachingscience.org*

CHAPTER 7

STUDENTS FROM RACIAL AND ETHNIC GROUPS AND THE *NEXT GENERATION SCIENCE STANDARDS*

MEMBERS OF THE *NGSS* DIVERSITY AND EQUITY TEAM

ABSTRACT

The student population in the United States is increasingly more racially and ethnically diverse. The U.S. Census collects data on race based on demographic groups of non-Hispanic white, Hispanic, African American, American Indian or Alaska Native, Asian, and Native Hawaiian or Other Pacific Islander. Although the National Assessment of Educational Progress (NAEP) science scores have improved in recent years, significant achievement gaps by racial subgroups persist. Current policies address these shortfalls by implementing guidelines for tracking the progress of underrepresented groups of students in science and by calling to address science achievement gaps. Effective strategies for students from major racial and ethnic groups are categorized as follows: (1) culturally relevant pedagogy, (2) community involvement and social activism, (3) multiple representation and multimodal experiences, and (4) school support systems including role models and mentors of similar racial or ethnic backgrounds. The vignette below illustrates effective science teaching for students from major racial and ethnic groups as they engage in the *NGSS*.

VIGNETTE: CONSTRUCTING EXPLANATIONS TO COMPARE THE CYCLE OF MATTER AND THE FLOW OF ENERGY THROUGH LOCAL ECOSYSTEMS

While the vignette presents real classroom experiences of the *NGSS* implementation with diverse student groups, some considerations should be kept in mind. First, for the purpose of illustration only, the vignette is focused on a limited number of performance expectations. It should not be viewed as showing all instruction necessary to prepare students to fully understand these performance expectations. Neither does it indicate that the performance expectations should be taught one at a time. Second, science instruction should take into account that student understanding builds over time and that some topics or

CHAPTER 7

ideas require extended revisiting through the course of a year. Performance expectations will be realized by using coherent connections among disciplinary core ideas, science and engineering practices, and crosscutting concepts within the *NGSS*. Finally, the vignette is intended to illustrate specific contexts. It is not meant to imply that students fit solely into one demographic subgroup, but rather it is intended to illustrate practical strategies to engage all students in the *NGSS*.

INTRODUCTION

Sequoyah Middle School is a large grade 6–8 urban school. Many of the students take the city bus to school because they live more than a mile away. Students from a government-subsidized housing neighborhood on the edge of the city also bus to Sequoyah because their neighborhood has no middle school. Another group of students come from the small residential neighborhood surrounding the school. Sequoyah is 65% nonwhite. In the vignette, 20 of the 29 students are from major racial or ethnic groups as described by the U.S. Census.

The teacher, Ms. C., demonstrates effective teaching strategies that motivate her students to participate in the science learning community and engage in the *NGSS* three-dimensional learning. Ms. C. is well loved by students because of her nurturing personality. The students know they are welcome in her classroom and often drop in for help before and after school. Despite Ms. C.'s easygoing nature, she insists on maintaining high expectations for all students. She knows that being scientifically literate means that her students not only understand science, but also form connections to the impact that science has on their lives. In the vignette, Ms. C. makes a point of connecting the students' community and real-world issues with a natural phenomenon and disciplinary core ideas. The life cycle assessment "walk-through" is based on the unit from the Great Lakes Bioenergy Research Center (2007–2012). Throughout the vignette, classroom strategies that are effective for all students, particularly for students from diverse racial and ethnic groups, are highlighted in parentheses.

RACIAL AND ETHNIC CONNECTIONS

Building on Students' Background Knowledge

Many of Ms. C.'s students studied the carbon cycle in seventh grade, but considering the high rate of student mobility, she couldn't take for granted that every eighth-grade student had mastered this challenging concept. For that reason, she reviewed photosynthesis the week before starting her alternative energy unit. Together, they had constructed explanations for the role of sunlight as the energy source that plants use to produce sugars from carbon dioxide and water.

This week Ms. C. began teaching alternative energy, and by the second day, Ms. C. had already had an invited speaker and completed an interactive multimedia presentation

Students From Racial and Ethnic Groups and the *Next Generation Science Standards*

to review the carbon cycle. She used these experiences as a springboard for a discussion about alternative fuels. *(Ms. C. used technology to present information in multiple modes of representation.)* She handed out corn chips to her surprised and delighted third-hour, eighth-grade students.

Every one of Ms. C.'s five classes had its own unique character. Her third-hour class was especially dynamic, and she looked forward to their fast-paced interchanges each day. Ms. C. had worked hard to engender a respectful atmosphere for discussion, and the 29 students had a relaxed rapport that reflected their respect for each other's individuality. Hand-raising was not required in Ms. C.'s room. She had come to realize, over the years, that a lack of formality encouraged participation from those less likely to contribute in traditional settings. She used creative ways to foster individual responsibility for learning, such as picking names (freedom sticks), share-outs, and cooperative grouping. To reinforce personal and collective responsibility in cooperative groups, Ms. C. employed a technique called roll-a-number. She assigned numbers 1 to 6 to the members of each team and then solicited arguments from the team members whose number was rolled (Hollie 2011).

Ms. C.'s classes had students from many different racial and ethnic backgrounds, reflecting the overall makeup of the middle school. She continually reinforced the idea that scientific discussions are more robust when there are lots of different perspectives and that each student has the responsibility to enrich the group's knowledge by contributing her or his perspective.

The day before, an ecologist and friend of Ms. C. who had just returned from a global conservation exchange in Nigeria presented a slide show about two countries in Africa. The ecologist described some social and environmental impacts of the oil industry on Nigerian ecosystems. He showed photos depicting the devastation of oil spills on the aquatic system. The class was clearly captivated by his presentation. Many had not previously considered where oil came from.

Bisma crunched on her chips and shared her memory of oil rigs in the Middle East. "They completely cover the land," she said, "Everywhere you look, it looks like those ducks that go into the water. What are they called? Dipping ducks?"

Ivy, a Nigerian American, had listened carefully to the presentation. She said that her family had come to the United States because of their religion, and she had never been told about the oil companies.

Nayeli added her experience with oil rigs: "I know what you mean! The air has a smell; it's like heavy. Sometimes I can feel oil in the air. The oil in the air *sticks* to you, and it messes with your hair." Nayeli had recently moved from Texas.

Some other students were discussing a new piece of information they had learned in today's multimedia presentation—lip gloss is made from petroleum products.

Ms. C. generously passed around salsa to her hungry class, asking, "Why do you think I brought in the *corn* chips to go with my PowerPoint?" After a few joking comments, Novid

CHAPTER 7

responded, "We are going to talk about corn in gas. I think I saw an ad where they put corn oil or something in their gas, ethanol. It is like corn. They said that the ethanol was better for the environment."

Delonné was surprised by that. "I like to pump the gas sometimes, and I saw that ethanol sign," she said. "I didn't know that the ethanol came from corn, why *corn*?"

Sarah answered, "It's cheaper." Delonné examined her corn chip and made a face, and a few students laughed. It *was* really weird to think about.

Ms. C. told the students that corn was subsidized by the government, which made corn very inexpensive and also resulted in high fructose corn syrup being added to lots of food. Suddenly a lot of students had something to say, so Ms. C. had to revert to calling on raised hands. Students shared concerns about corn being used for fuel when people need it to eat, and high fructose corn syrup being unhealthy. Many students suggested that other alternative fuels are better because they are not sources of food. Ms. C. was impressed that students knew alternative ways to make fuels, such as solar energy, wind energy, water energy, and even geothermal energy. She wrote them all down on a list to refer to throughout the unit.

Ms. C. assigned two open-ended homework assignments: to think about where some of these different fuels came from and to generate a list of food items in their homes that had corn as one of the ingredients. Also, as part of the weekly assignment, she asked each student to look up one alternative fuel that was developed or was being developed locally and to write down facts about what they learned.

After 13 years of teaching, Ms. C. had learned to capitalize on the joint familiarity of the neighborhood and school as a bridge to form common understandings. She believed that place-based pedagogy was an effective way to reach her students since they came from diverse cultural backgrounds. (*Ms. C. used culturally relevant pedagogy by connecting the science curriculum to the students' cultural experiences.*)

Definition of Efficiency

The next day, Delonné had an announcement to make to the class. She told the class that she had pumped gas for her mother's car yesterday and that the pump had said 10% ethanol. She was very excited to report this. She also reiterated her concerns about taking corn out of production as a food source. She thought that the price of food would go up if too much corn was used for fuel, especially considering how many foods contained corn. Delonné was an African American student who maintained a high level of participation in class. One of the other classes in Delonné's schedule was AVID (Advancement Via Individual Determination), a program intended to support college-bound minority and economically disadvantaged students in taking higher-level classes. (*AVID is a school-based support system for students of diverse racial and ethnic groups.*)

Students From Racial and Ethnic Groups and the *Next Generation Science Standards*

Several students brought in lists of foods in their homes that contained corn. Amira, another AVID student, brought in a list of 50 items and stated she was shocked at the number of foods that contained corn. Madison found only four items in her home because her mother made a conscious effort not to buy food with corn or corn products added. Marcus brought several packages of candy with him as evidence that all candy contained corn syrup. Tom confirmed Marcus's statement. Tom was a second generation Mexican American who rarely spoke in the whole group, but he was a leader on small teams. Many students, following Delonné's lead, voiced their concerns about rising food costs if corn is used as fuel. Novid, however, firmly stated, "It's better for the environment." Not easily intimidated by dissent, Novid, a Nepai American, held his ground during class discussions.

Ms. C., enjoying the animated discussion, jumped in and seized on the last comment that corresponded to her learning goal for the day: "Why do people say that some fuels are better for the environment? Are some fuels more efficient? How could we decide?" She wrote the driving question on the interactive whiteboard: "Are some fuels more efficient than others?" The purpose of the activity was for students to explore their definitions of *efficiency*.

Ms. C. distributed the biofuels preassessment questionnaire, *Thinking About Plants as Transportation Fuel* (GLBRC 2007–2012). Students discussed the questions from the sheet in their self-selected teams. The questionnaire aimed at sparking discussions about types of fuels, their carbon output, impact on the environment, and energy to produce the fuels. Also, the questionnaire directed the self-selected teams to represent their own simplified carbon cycle model to explain the production of ethanol from plant biomass, using pictures and words (practice: Developing and Using Models). This model would be revisited repeatedly as the students improved their understanding of matter and energy, and the student teams used the model to record thinking, test new claims, and ask new questions.

Next, small teams of three were partnered up to share their thoughts. Walking among the groups, Ms. C.'s biggest concern was one of categorization: Students confused *matter* with *energy*, or used *gasoline* and *energy* interchangeably. For example, one group wrote in their narrative under their drawing, "Plant biomass goes to energy from the plant and becomes ethanol, then goes to the refinery, then to gas station, then to car." They had acquired some appropriate terms for the process, but needed to develop a greater understanding of the complicated set of chemical reactions. A few students attempted to use chemical equations to demonstrate their understanding of photosynthesis and combustion in a car engine.

Interesting discussions emerged from the "true or false" statement "Creating ethanol from plant biomass contributes to climate change." Novid was resolute that ethanol was not harmful to the environment. However Andrés insisted, "If it comes out of a car, then the carbon is definitely harmful carbon."

The class unanimously agreed that a person could judge the efficiency of a fuel based on the mileage compared to the amount of released carbon into the environment when "burned." But Bisma commented, "Even though gas is more efficient than ethanol because it gets better miles per gallon, it isn't good for other countries, like Nigeria. Efficiency isn't the only important thing!" Some students nodded. Ms. C. asked, "Can our definition of *efficiency* include impacts of production?" After a long pause, she wrote down the preliminary definition of *efficiency* and told the class that it would be revised as they learned more: "*Efficiency* means: how many miles per gallon a fuel gets compared to the carbon emissions per gallon." She added Bisma's comment about considering other factors in the definition of *efficiency* on the interactive whiteboard.

At the close of the hour, Ms. C. asked students to consider alternative energy options by asking questions that led to further investigations (practice: Asking Questions and Defining Problems). Jarek asked a question that he had been pondering: "Ms. C., in your presentation you said that the energy for gas is from when the sugar from the plant reacts with oxygen. Could that be true for any plant, not just corn?" Ms. C. nodded encouragingly. Jarek offered an idea that showed his developing understanding of energy, "We should just collect grass and turn that into fuel—everyone is talking about needing corn for food, but nobody really eats grass!"

Revising the Model and Constructing Explanations

For a few days, students read, discussed, and presented articles and preliminary research about ethanol and other alternative fuels. They also observed the products of a "burning candle" using a projector camera. Next, teams worked together to participate in a "walk-through"—an initial model for the life cycle of the matter in ethanol. In the "walk-through" the students themselves acted out the life cycle assessment of ethanol. The students held cards with names of molecules on them and used yarn to track their "journey." They had already completed this modeling activity one time through, representing matter transformation with purple yarn. Ms. C. decided to return to it, address lingering questions about matter, and consider the flow of energy this time.

Some students were confusing the words *matter*, *fuel*, and *energy*, and were using them interchangeably. They were unclear about the relationship between energy and matter throughout the process. Although energy can be traced similarly to matter in the ecosystem, she needed the students to see how energy is fundamentally different from matter in basic ways. Through understanding this process, students would be able to see that the transfer of energy drives the cycling of matter, and the transfer of energy can be tracked as energy flows through a natural system (CC: Energy and Matter; DCI: MS-LS2D Ecosystems, PS1.B Matter and Its Interactions).

As the eighth-grade class streamed into the science classroom and found their seats at the tables, Ms. C. opened the file with names of the week's project teams on the interactive whiteboard. She believed in randomly rearranging team groupings, giving her students the opportunity to interact and learn from both self-selected and dissimilar sets of peers.

Students From Racial and Ethnic Groups and the *Next Generation Science Standards*

She fostered a vital social support system for students through maintaining self-selected partnerships. She used the interactive whiteboard application to combine the smaller self-selected teams of two or three with one other team.

Ms. C. posted three now-familiar ethanol production "stations" on the classroom walls, marked by large printed signs and homemade blown-up pictures of local areas: (1) a farm field, (2) a refinery, and (3) the gas station near the school. She had a copy of the Life Cycle Assessment Process Tool (based on the Matter and Energy Process Tool developed by Charles Anderson and colleagues. of the Environmental Literacy Project) and planned to model matter and energy separately to emphasize the idea that matter could be traced separately from energy. These stations represented key points for inputs, outputs, and transformation of matter and energy as crops are grown, converted into, and used as fuel. The class had learned through research that ethanol production is a fermentation process—the microbes (yeast) digest the sugar and produce ethanol and carbon dioxide (CO_2) as by-products (practice: Obtaining, Evaluating, and Communicating Information). This activity would help students test their understanding of how matter is cycled by living and nonliving components of an ecosystem through the process of fermentation, as well as the overall life cycle of a biofuel.

As the class talked through the Life Cycle Assessment Process Tool, Ms. C. used the same photos from the stations on the interactive whiteboard, drew three rectangles, and wrote their designations: farm, refinery, and car. She drew straight purple lines to indicate matter input and output, and wavy red lines to indicate energy input and output. She had cut purple and red yarn for students to represent and track the journey that matter took throughout its life cycle.

Ms. C. announced, "Last week, we traced matter input and output for the carbon cycle of ethanol production and combustion, and we had some unanswered questions. Today, we are going to see where energy fits into the process." (*Ms. C. chose a class modeling activity that involved student movement, a strategy that uses a multimodal experience to increase student engagement.*)

She laid out the matter cards faceup on the table, using the document camera, so the class could read them as she laid them out. Each card had a word (e.g., water) and the chemical formula (e.g., H_2O) in clear black letters. "We are at the farm. Who will be matter input for the farm? Michael, you are Carbon Dioxide (CO_2) input for photosynthesis, okay?" Ms. C. used the interactive whiteboard application to help her randomly choose students for this activity. Michael (CO_2) and Kayla (H_2O) together found the Carbon Dioxide and Water cards on the table, and two pieces of purple yarn. Then they each gave one end of their yarn to Axel, who was standing under the farm picture ready with his Glucose card.

The yarn illustrated the tracing of matter in a concrete way, representing the path that matter takes to create corn biomass. Ms. C. suspected that using another color yarn for energy would clear up the confusion many students were having. The class resumed

CHAPTER 7

tracing the plant biomass matter to the refinery, with Delonné holding another Glucose card and the other end of Axel's yarn under the refinery station. Then the matter path led to the car at the gas station, where Ivy held onto the other end of Axel's yarn and an Ethanol card. Finally, the path resumed to show that the ethanol created from the plant biomass reacted with oxygen to form carbon dioxide, Austin's card, as the fuel was burned in the car.

Marcus still held on to the one H_2O matter card. His team had not been sure where it went. They placed their unresolved question in the question board: "Does H_2O go with the refinery or the farm—as an output—or does it go as output for the car?" Ms. C. rewrote the question on the interactive whiteboard screen. Each of the three students in Marcus's team had made convincing arguments for water to go in all three stations.

Ms. C. had been looking forward to this discussion. "Did anyone have any more thoughts about the last Water card?" She looked pointedly at Shamaya, who had struggled with the carbon cycle and the life cycle of ethanol. This morning, she had told Ms. C. that she now "got it" and knew where the last H_2O went. She directed her words to Ms. C. and then to her team: "I've got this. It goes as an output of the car!" She was confident, saying, "There is steam that comes out of the car and you can see it. The smoke isn't carbon dioxide; it's steam—water!" Shamaya explained that she had made the connection the night before, "I used to think that the smoke that comes out of the car was carbon dioxide, but it isn't; it's steam." Doubtfully, Tom pointed out that the smoke that comes out of a car smells bad, and steam doesn't smell like that. But Shamaya was undaunted, "Carbon dioxide is invisible, so the smoke coming out of the car is maybe a mixture of carbon dioxide and steam. I know it seems funny, like it should be the 'anti-water,' but, like, burning the hot candle also made water." Shamaya was referring to evidence from the earlier burning candle investigation (practice: Engaging in Argument From Evidence).

Ms. C. prompted, "Say more about that, Shamaya." Shamaya said, "When you burned the candle and put the glass bowl over it, it made black smoke; that was carbon, I think, and it also made steam, uh, water drops that were on the glass bowl and carbon dioxide." Jarek nodded, agreeing with Shamaya, and added that if they put a glass bowl over the pipe from the car, it might make water drops too. He suggested that since the last outputs and the first inputs are the same molecules, carbon dioxide and water, the people holding those cards could actually be the same person: carbon dioxide and water make sugar and oxygen, and then the sugar reacts with oxygen to form carbon dioxide and water (Figure 7.1; practice: Developing and Using Models).

Ms. C. realized that Shamaya and Jarek were still working through their understanding of the processes at hand, but felt certain that this would be resolved as the unit progressed. What they were doing—connecting the cycling of matter to the processes of photosynthesis and combustion of materials—was helping the students construct an explanation about the cycle of matter (practice: Constructing Explanations and Designing Solutions).

Students From Racial and Ethnic Groups and the *Next Generation Science Standards*

FIGURE 7.1.

WALK-THROUGH

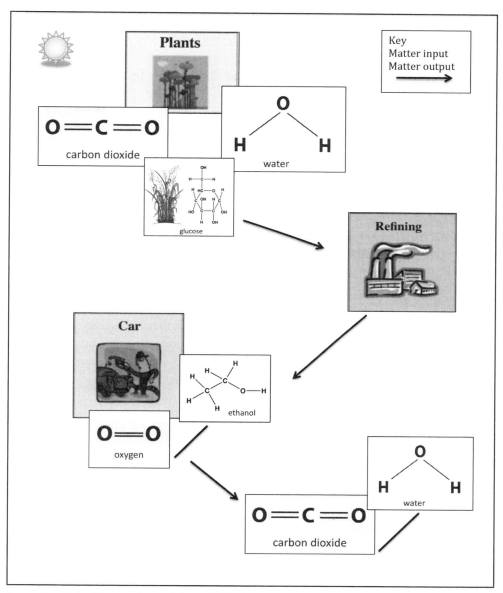

Source: Great Lakes Bioenergy Research Center 2007–2012.

CHAPTER 7

After checking for agreement and some negotiation, Marcus took his place under the picture of the car. To complete the circle, Marcus, the H_2O output from the car, gave an end of his second piece of yarn to Kayla, the H_2O input at the farm. They were satisfied that a spider web of yarn made an obvious circle from one station to the next, around the room. Following suit, Michael, the carbon dioxide farm input, and Austin, the carbon dioxide car output, also connected their yarn and closed the yarn circle.

Comparing the Cycle of Matter to the Flow of Energy in the Carbon Cycle

The students had modeled the cycle of matter for ethanol, but now they needed to use red yarn to add the flow of energy.

Before more students got out of their seats, Ms. C. used the same stations (station 1: farm field, station 2: refinery, station 3: car) to discuss the energy involved. Considering and including the energy flow for the carbon cycle of ethanol was challenging but invigorating for the class (DCI: MS-LS2D Ecosystems; PS1.B Matter and Its Interactions).

For station 1, the farm, every group very confidently named sunlight as the energy input. One team was unanimous that the energy was located in the bonds within the cellulose, starch, and sugar. Ms. C. was quick to remind the students that energy was in the system, not exactly in the bonds. She held up a slack rubber band for a model, saying, "There is no energy in the rubber band until I stretch it." She had Tom pull the other end of the rubber band: "Now there is energy in the system."

There was some disagreement among students as to whether any energy was lost during the first process of photosynthesis. Ms. C. added the question to the question board. At each step in the *Life Cycle Assessment Process Tool*, Ms. C. had groups of 4–6 students discuss their thoughts and commit to a decision. For students to share out with the larger group, Ms. C. used her engagement protocol, roll-a-number, and called out a team member's number to pick one person from each team to share the team's thoughts. This was how she communicated her expectation for individual responsibility from each team member.

At station 2, the refinery, the question of input and output of energy in fermentation of the plant-derived sugars resulted in a debate. Teams thought the plant product would need to be cut up, cooked or chopped up, and that would definitely take some energy. Nayeli reminded the class about one of the articles they had read. She said that enzymes are added at the refinery to start breaking down the matter, like the enzymes in your saliva. Most heads nodded. The question about energy output at this station was written on the board for further investigation.

Sorting out the process of combustion resulted in another interesting discussion. Students disagreed about the energy output during combustion. Michael said, "The molecules break apart, and then they go into the air, the energy was in the system … and it is burned up." Novid said, "What do you think the car does right after we put gas into it? The car goes!" Ms. C. asked students to think about other outputs that might be taking place during the

Students From Racial and Ethnic Groups and
the *Next Generation Science Standards*

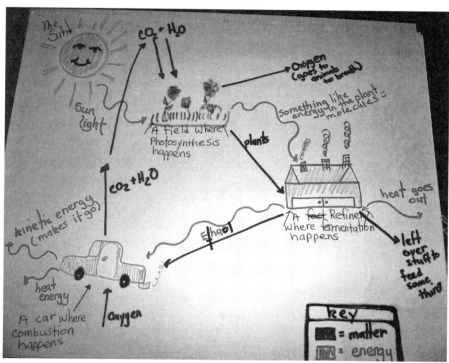

FIGURE 7.2.

STUDENT DRAWING OF THE CYCLE OF MATTER AND FLOW OF ENERGY

last stage, asking "Does a running car produce heat? Does it produce any sound?" She was comfortable leaving some questions unanswered for the time being, so that she could pursue them throughout the course of the unit.

Application of Core Ideas

In the subsequent lesson, Ms. C. completed the walk-through activity using red energy yarn to trace the flow of energy. The students used the model to further explore the relationship between matter and energy (CC: Energy and Matter). Ms. C. needed the students to grasp the idea that matter is conserved and so is energy, although energy is transferred throughout the process (Figure 7.2). Ms. C. ended the unit with the students working in teams to predict the cycle of matter and the flow of energy for ethanol made from switchgrass (an alternative fuel, from native prairie grass, being investigated in their state) and

FIGURE 7.3.

STUDENTS DEFENDING THEIR MODELS USING SCIENTIFIC ARGUMENTATION

defending their models with scientific argumentation (Figure 7.3; practice: Engaging in Argument From Evidence).

The students worked together to refine their list of important factors for evaluating efficiency and usability when comparing fuels. This required more research about the broad-scale and local impacts of the use of corn as a fuel (ETS1.B: Developing Possible Solutions; ETS1.C: Optimizing Design Solutions).

Ms. C. had the class read a newspaper article about a proposal in the state budget to use local power plants and biomass from the state's agricultural fields to create energy. Each student applied his or her new science understandings and composed a letter to a local representative supporting or opposing the proposal. (*Ms. C. connected science to locally relevant issues that promote community involvement and social activism*) (practice: Obtaining, Evaluating, and Communicating Information).

Students From Racial and Ethnic Groups and the *Next Generation Science Standards*

Finally, Ms. C. asked her students to put together presentations showing what they found most meaningful in the unit. Delonné wrote a rap about the carbon cycle, comparing ethanol and gasoline. The students loved it. Every day Marcus and Axel sang the first few lines of Delonné's rap to Ms. C. when they walked to their lockers. Another group wrote a short play and another put together a mock panel discussion. These shared experiences of music and media, part of youth culture, made the science content engaging to the students. This assignment offered opportunities for literacy integration and cultural relevance. (*The teacher used culturally relevant pedagogy by connecting the science curriculum to the students' cultural experiences.*)

At the end of the unit, Ms. C. asked her students to write how the unit was important for them and how this lesson connected to their lives.

Delonné wrote, "This is so important to me, because when I am an adult, in my future there might be a crisis. And all the countries might run out of oil and all I will have left to put in my car is corn and switchgrass. Our parents are going to be gone soon and we will have to fix it. We have to come up with a solution to this!"

Novid explained, "This is going to affect my future because we use energy every day and we don't want the Earth to get so polluted. Ethanol may be a better solution to keep our Earth clean. Almost every kind of energy has some problems with it; we can still pick one that is the most efficient. Scientists can make better enzymes and protect what we have left."

Andrés wrote, "My dad needs gas for his car and uses ethanol in it. Then the car releases CO_2 and water in the air. It affects the planet and our world will change. We could have tornadoes and earthquakes."

And Marcus wrote, "This connects to my life because I want to know my future and what adults have done wrong, so when we get older we can change it. And we need to know more about ethanol and gasoline and switchgrass because I want to drive my parents' cars and other people's cars. It is going to affect me because all of it is going to affect the environment I will live in for the rest of my life."

NGSS CONNECTIONS

The *NGSS* aim to provide every student with a more comprehensive understanding of science by blending disciplinary core ideas with science and engineering practices and crosscutting concepts. This new approach allows student participation in science to more closely reflect what scientists and engineers do in the real world. The teacher in the vignette, Ms. C., teaches students from diverse racial and ethnic backgrounds by employing effective strategies that enable the students to meet the standards. See Figure 7.4 (p. 79) for the comprehensive list of *NGSS* and *CCSS* from the vignette.

In the vignette, the students developed an understanding of the three dimensions to meet the middle school grade-band performance expectations in life sciences (LS1: From

CHAPTER 7

Molecules to Organisms: Structures and Processes) and physical sciences (PS1: Matter and Its Interactions) and were introduced to some core ideas in ETS1: Engineering Design.

Performance Expectations

MS-LS2-3 Ecosystems: Interactions, Energy, and Dynamics
Develop a model to describe the cycling of matter and the flow of energy among living and nonliving parts of an ecosystem.
MS-PS1-3 Matter and Its Interactions
Gather and make sense of information to describe that synthetic materials come from natural resources and impact society.
MS-ESS3-3 Earth and Human Activity
Apply scientific principles to design a method for monitoring and minimizing a human impact on the environment.
MS-ESS3-4 Earth and Human Activity
Construct an argument supported by evidence for how increases in human population and per-capita consumption of natural resources impact Earth's resources.

Disciplinary Core Ideas

LS2.B Cycles of Matter and Energy Transfer in Ecosystems

Transfers of matter into and out of the physical environment occur at every level. For example, when molecules from food react with oxygen captured from the environment, the carbon dioxide and water thus produced are transferred back to the environment, and ultimately so are waste products, such as fecal material. The atoms that make up the organisms in an ecosystem are cycled repeatedly between the living and nonliving parts of the ecosystem.

PS1.B Chemical Reactions

Substances react chemically in characteristic ways. In a chemical process, the atoms that make up the original substances are regrouped into different molecules, and these new substances have different properties from those of the reactant.

ETS1.B Developing Possible Solutions

There are systematic processes for evaluating solutions with respect to how well they meet the criteria and constraints of a problem.

ETS1.C Optimizing the Design Solution

The iterative process of testing the most promising solutions and modifying what is proposed on the basis of the test results leads to greater refinement and ultimately to an optimal solution.

CHAPTER 7

Science and Engineering Practices

Developing and Using Models

Develop a model to predict and/or describe phenomena.

Engaging in Argument From Evidence

Respectfully provide and receive critiques about one's explanations, procedures, models, and questions by citing relevant evidence and posing and responding to questions that elicit pertinent elaboration and detail.

Obtaining, Evaluating, and Communicating Information

Gather, read, and synthesize information from multiple appropriate sources and assess the credibility, accuracy, and possible bias of each publication and methods used, and describe how they are supported or not supported by evidence.

Constructing Explanations and Designing Solutions

Construct an explanation using models or representations.

Throughout the unit, the students developed mastery of sciennce practices by authentically engaging with scientific ideas. Among the many science practices the students engaged with, only a few were specifically targeted in this vignette. Students *engaged in argument* regarding the validity of their scientific explanations, and incorporated their new thinking into their models. Students *obtained, evaluated, and communicated information* by researching alternative fuels and used some of the evidence they gathered to support their claims. Ms. C. supported her students in *developing and using models* to *construct explanations* about the life cycle of matter and energy in the ecosystem. Then the students analyzed their model collaboratively and refined it as new information came to light.

Students From Racial and Ethnic Groups and the *Next Generation Science Standards*

Crosscutting Concepts

> **Energy and Matter**
>
> The transfer of energy can be tracked as energy flows through a natural system.
>
> Within a natural or designed system, the transfer of energy drives the motion and/or cycling of matter.

The unit addressed learning outcomes for crosscutting concepts. Students used the crosscutting concept to understand that energy can be traced and that it is conserved. The students' understanding of the crosscutting concept of Energy and Matter was deepened through revisiting core ideas of energy and matter from earlier units in physical and life sciences, building on these core ideas, and reaching the grade level objective. Ms. C. selected driving questions that would facilitate cross-disciplinary connections to energy and matter.

CCSS CONNECTIONS TO ENGLISH LANGUAGE ARTS (ELA) AND MATHEMATICS

The *NGSS* promotes a vision of science learning as an interdisciplinary undertaking and each standard includes the *CCSS* connections to English language arts (ELA) and mathematics. The vignette highlights the dynamic integration of science with ELA and math standards to ensure student learning across disciplines.

As part of the Energy and Ecosystems unit, each student conducted research on alternative energy sources and presented their findings to the class. This research enhanced the students' mastery of the science content by using a community context, and the activity gave students added practice of research, writing, and presentation skills. The unit addressed the *CCSS ELA*:

- **W.6.8** *Gather relevant information from multiple print and digital sources; assess the credibility of each source; and quote or paraphrase the data and conclusions of others while avoiding plagiarism and providing basic bibliographic information for sources.*

The unit also incorporated the *CCSS Mathematics*. As the students compared the efficiency of various fuels, they discovered that more factors needed to be considered. Rather than just number of miles-per-gallon and carbon output, students also considered the carbon output of the heavy machinery used in production, the transportation and converting costs, and the social and environmental impacts, to name a few. For example, switchgrass, a native species of prairie grass, requires relatively little energy to harvest, yet the energy needed to transform the biomass to usable fuel is intensive. Students used mathematics to compare and contrast these data. The four *CCSS Mathematics* that are appropriate for middle school and that are met through the course of the unit are:

- **MP.2** *Reason abstractly and quantitatively.*
- **S.IC** *Make inferences and justify conclusions from sample surveys, experiments, and observational studies.*
- **MP.4** *Model with mathematics.*
- **5.OA** *Analyze patterns and relationships.*

EFFECTIVE STRATEGIES FROM RESEARCH LITERATURE

In addition to extensive literature about the importance of multicultural education in our increasingly diverse schools and classrooms, emerging literature presents classroom strategies specific to content areas, such as science. According to research, effective strategies for students from major racial and ethnic groups fall into the following categories: (1) culturally relevant pedagogy, (2) community involvement and social activism, (3) multiple representation and multimodal experiences, and (4) school support systems including role models and mentors of similar racial or ethnic backgrounds (Lee and Buxton 2010).

First, culturally relevant pedagogy (Ladson-Billings 1995) values and affirms the different cultures and backgrounds of the students in the classroom. The teacher integrates and connects the cultures of the students (including funds of knowledge, family and community histories, and linguistic backgrounds) to academic content and practices. Through culturally relevant pedagogy the teacher communicates high expectations that science is relevant and valuable for all students.

Second, community involvement and social activism in science involves creating meaningful, place-based lessons that may result in social activism. Connecting science concepts to the daily lives and futures of students from diverse backgrounds motivates them to learn science. Furthermore, they increase their participation in science when they see scientific knowledge as a way to gain entry to affect positive change in community contexts.

Third, multiple modes of representation include different forms of literacy, including print, technology, and media, as well as aural forms. Multimodal experiences allow students to use senses and modalities to absorb material, including tactile, kinesthetic, and acting-out scenarios.

Finally, school support systems are critically important for success in science achievement and careers for students of diverse races and ethnicities. School support systems include teachers, guidance counselors, mentors, and peers from the students' own backgrounds who support science learning. They provide examples of success in science and ways to navigate school systems. School support systems also reach out to students'

Students From Racial and Ethnic Groups and
the *Next Generation Science Standards*

FIGURE 7.4.

NGSS AND CCSS FROM VIGNETTE

MS-LS2 Ecosystems: Interactions, Energy, and Dynamics
MS-PS1 Matter and Its Interactions
MS-ESS3 Earth and Human Activity

Students who demonstrate understanding can:

MS-LS2-3. Develop a model to describe the cycling of matter and the flow of energy among living and nonliving parts of an ecosystem.

MS-PS1-3. Gather and make sense of information to describe that synthetic materials come from natural resources and impact society.

MS-ESS3-3. Apply scientific principles to design a method for monitoring and minimizing a human impact on the environment.

MS-ESS3-4. Construct an argument supported by evidence for how increases in human population and per-capita consumption of natural resources impact Earth's systems.

The performance expectations above were developed using the following elements from the NRC document *A Framework for K–12 Science Education*:

SCIENCE AND ENGINEERING PRACTICES	DISCIPLINARY CORE IDEAS	CROSSCUTTING CONCEPTS
Developing and Using Models Modeling in 6–8 builds on K–5 and progresses to developing, using and revising models to support explanations, describe, test, and predict more abstract phenomena and design systems. • Develop a model to describe phenomena **Obtaining, Evaluating, and Communicating Information** Obtaining, evaluating, and communicating information in 6–8 builds on K–5 and progresses to evaluating the merit and validity of ideas and methods. • Gather, read, and synthesize information from multiple appropriate sources and assess the credibility, accuracy, and possible bias of each publication and methods used, and describe how they are supported or not supported by evidence. **Engaging in Argument From Evidence** Engaging in argument from evidence in 6–8 builds on K–5 experiences and progresses to constructing a convincing argument that supports or refutes claims for either explanations or solutions about the natural and designed world. • Evaluate competing design solutions based on jointly developed and agreed-upon design criteria.	**LS2.B: Cycle of Matter and Energy Transfer in Ecosystems** • Food webs are models that demonstrate how matter and energy is transferred between producers; consumers, and decomposers as the three groups interact within an ecosystem. Transfers of matter into and out of the physical environment occur at every level. Decomposers recycle nutrients from dead plant or animal matter back to the soil in terrestrial environments or to the water in aquatic environments. The atoms that make up the organisms in an ecosystem are cycled repeatedly between the living and nonliving parts of the ecosystem. **PS1.B: Chemical Reactions** • Substances react chemically in characteristic ways. In a chemical process, the atoms that make up the original substances are regrouped into different molecules, and these new substances have different properties from those of the reactants. **ETS1.B: Developing Possible Solutions** • There are systematic processes for evaluating solutions with respect to how well they meet the criteria and constraints of a problem. **ETS1.C: Optimizing the Design Solution** • The iterative process of testing the most promising solutions and modifying what is proposed on the basis of the test results leads to greater refinement and ultimately to an optimal solution.	**Energy and Matter** • The transfer of energy can be tracked as energy flows through a natural system. • Within a natural or designed system, the transfer of energy drives the motion and/or cycling of matter.

CCSS **Connections for English Language Arts and Mathematics**

W.6.8 Gather relevant information from multiple print and digital sources; assess the credibility of each source; and quote or paraphrase the data and conclusions of others while avoiding plagiarism and providing basic bibliographic information for sources.

MP.2 Reason abstractly and quantitatively.

S.IC Make inferences and justify conclusions from sample surveys, experiments, and observational studies.

MP.4 Model with mathematics.

5.OA Analyze patterns and relationships.

CHAPTER 7

families and communities to address the important role they play in supporting the students. A key factor for students choosing a career in science is contact with successful, caring scientists from their own background.

CONTEXT

DEMOGRAPHICS

According to the 2010 U.S. Census, 36% of the U.S. population is composed of racial minorities, of whom 16% are Hispanic, 13% African American, 5% Asian, and 1% American Indian or Native Alaskans (U.S. Census Bureau 2012). The student population in the United States is increasingly more diverse racially and ethnically. Forty-five percent of the school age population under 19 years old was a racial minority in 2010. It is projected that the year 2022 will be the turning point when minorities will become the majority in terms of percentage of the school-age population.

As an indication of changing demographics, the National Assessment of Educational Progress (NAEP) report card for racial and ethnic categories beginning in 2011 consists of these demographic categories: White, African American, Hispanic, Asian, Native Hawaiian/Other Pacific Islander, American Indian/Alaska Native, and two or more races (NCES 2012, p. 7).

SCIENCE ACHIEVEMENT

While the student population in the United States is becoming more diverse, science achievement gaps have persisted by demographic subgroups. At the national level, NAEP provides assessment of U.S. students' science performance that shows wide achievement gaps among racial groups. For example, on the most recent 2011 NAEP Science assessment administered to eighth grade only, the average score for African American students was 129, for Hispanic students was 137, and for white students was 163 (NCES 2012). The 2011 science performance also points to alarming underrepresentation of racial minority students in the Advanced Proficient (above 75%) category and overrepresentation in the Below Proficient (below 25%) category. In the Advanced Proficient category, 76% of the students were white, 10% Hispanic, and 4% African American. This trend reversed in the Below Proficient category, as 35% were Hispanic, 31% African American, and 27% white. Furthermore, science achievement gaps among racial groups persist over time, consistent at grades 4, 8, and 12.

EDUCATION POLICY

The Elementary and Secondary Education Act of 2001 (ESEA) says that adequate yearly progress (AYP) should apply the same standard to all students in a state, with "separate

Students From Racial and Ethnic Groups and the *Next Generation Science Standards*

measurable annual objectives for continuous and substantial improvement in each category including students from major racial and ethnic groups" at the 95% achievement level (Title I Part A Subpart 1 Sect 1111. (b)(2)(I)(ii)). The disaggregation would not be required in a case where the numbers in a category are statistically not significant or where the identity of an individual student would be revealed.

> *Title I is intended for "improving the academic achievement of the disadvantaged" in order to meet "the educational needs of low-achieving children in our Nation's highest-poverty schools, limited English proficient children, migratory children, children with disabilities, Indian children, neglected or delinquent children, and young children in need of reading assistance." Part A states that, beginning no later than school year 2005–2006, the states were required to measure the proficiency of all students in science (Title I Part A Subpart 1 Sec. 1111 (b)(1) (C)) not less than one time during: grades 3–5, grades 6–9, and grades 10–12 (Title I Part A Subpart 1 Sec. 1111 (b)(3) (C) (II)) ESEA calls for educational agencies hold American Indian and Alaskan Native students to the same "challenging State student academic achievement standards as all other students are expected to meet."* (ESEA, 2001 Title VII Sec. 701 Part A. Sec. 7102 (a))

In addition, the Race to the Top Priority 2: Competitive Preference Priority–Emphasis on Science, Technology, Engineering, and Mathematics (STEM) (U.S. Deptartment of Education 2009) is calling for schools to "prepare more students for advanced study and careers in the sciences, technology, engineering, and mathematics, including by addressing the needs of underrepresented groups and of women and girls in the areas of science, technology, engineering, and mathematics."

REFERENCES

Anderson, C., et al. 2010. Environmental literacy project: Carbon cycle. East Lansing, MI: Michigan State University. *http://envlit.educ.msu.edu/publicsite/html/carbon.html*

Elementary and Secondary Education Act of 1965, Pub. L. No. 89–10, 79 Stat. 27.

Great Lakes Bioenergy Research Center. 2007–2012. Life cycle assessment of biofuels 101. *www.glbrc.org/education/educationalmaterials/biofuels-101*

Hollie, S. 2011. *Culturally and linguistically responsive teaching and learning: Classroom practices for student success.* Huntington Beach, CA: Shell Education Press.

Ladson-Billings, G. 1995. Toward a theory of culturally relevant pedagogy. *American Educational Research Journal* 32: 465–491.

Lee, O., and C. A. Buxton. 2010. *Diversity and equity in science education: Theory, research, and practice.* New York: Teachers College Press.

National Center for Education Statistics (NCES). 2012. *The nation's report card: Science 2011* (NCES 2012–465). Washington, DC: U.S. Department of Education, Institute of Education Sciences. *http://nces.ed.gov/nationsreportcard/pdf/main2011/2012465.pdf*

CHAPTER 7

U.S. Census Bureau. 2012. *Statistical abstract of the United States, 2012.* Washington, DC: U.S. Government Printing Office. *www.census.gov/compendia/statab/cats/education.html*

U.S. Department of Education. 2009. Race to the top program executive summary. Washington, DC: U.S. Department of Education.

CHAPTER 8

STUDENTS WITH DISABILITIES AND THE *NEXT GENERATION SCIENCE STANDARDS*

MEMBERS OF THE *NGSS* DIVERSITY AND EQUITY TEAM

ABSTRACT

The percentage of students identified with disabilities in schools across the nation is currently around 13%. As a result of the Elementary and Secondary Education Act (ESEA), school districts are held accountable for the performance of students with disabilities on state assessments. Although students with disabilities are provided accommodations and modifications when assessed, as specified in their Individualized Education Plans (IEP), achievement gaps persist between their science proficiency and the science proficiency of students without disabilities. The vignette below highlights effective strategies for students with disabilities: (1) multiple means of representation, (2) multiple means of action and expression, and (3) multiple means of engagement. These strategies support all students' understanding of disciplinary core ideas, science and engineering practices, and crosscutting concepts as described by the *Next Generation Science Standards* (*NGSS*).

VIGNETTE: USING MODELS OF SPACE SYSTEMS TO DESCRIBE PATTERNS

While the vignette presents real classroom experiences of *NGSS* implementation with diverse student groups, some considerations should be kept in mind. First, for the purpose of illustration only, the vignette is focused on a limited number of performance expectations. It should not be viewed as showing all instruction necessary to prepare students to fully understand these performance expectations. Neither does it indicate that the performance expectations should be taught one at a time. Second, science instruction should take into account that student understanding builds over time and that some topics or ideas require extended revisiting through the course of a year. Performance expectations will be realized by using coherent connections among disciplinary core ideas, science and engineering practices, and crosscutting concepts within the *NGSS*. Finally, the vignette is intended to illustrate specific contexts. It is not meant to imply that students fit solely into one demographic subgroup, but rather it is intended to illustrate practical strategies to engage all students in the *NGSS*.

INTRODUCTION

There are five sixth-grade classes at Maple Grove, the only middle school in a small rural school district. Approximately 10% of the K–12 school population receives special education services. The school has about 480 students in grades 6–8. The district population consists of 1,320 students: 92.3% white, 3.6% African American, 2% Hispanic, 0.5% Asian, and 0.3% Native American; 34% are classified as low socioeconomic status.

The incidence rates of identified special education students in the district are highest in the categories of specific learning disabilities (2.4%) and other health impairments including ADD/ADHD (2.7%). In addition, 1.1% of students are in the category of "speech impaired," 1.4% "language impaired," 0.8% "intellectual disabilities," and 0.8% "autism."

There are special education students in each of the sixth-grade classes with Individualized Education Plans (IEPs) that specify the accommodations and modifications when participating in the regular education classroom. Mr. O. thinks about potential barriers that any of his students, including those with special needs, may have to the planned instruction. Then he adjusts instruction to overcome those barriers. Often, changing an approach to accommodate barriers makes instruction more effective for all students. The students with disabilities, along with their regular education peers, receive science instruction from the science teacher five days a week for 50 minutes each day. Most of the identified students receive instruction in reading/language arts and mathematics in a coteaching model. Some students receive additional pullout services in those content areas or in social skills.

In the lesson sequence in this vignette, Mr. O. uses multiple means of representations for Moon phases—Stellarium (planetarium software), Styrofoam balls, a lamp, golf balls, and foldables (three-dimensional interactive graphic representations developed by Zike). Mr. O. provides additional practice for students who may need it, such as placing cards with Moon phases in chronological order and then identifying each phase. He modifies assignments for students with intellectual disabilities as mandated by their IEPs. In addition, strategic grouping of students provides support for struggling students, including special education students. Throughout the vignette, classroom strategies that are effective for all students, particularly for students with disabilities according to the research literature, are highlighted in parentheses.

SPECIAL EDUCATION CONNECTIONS

Jeanette and Nicole have intellectual disabilities; they have a paraprofessional who accompanies them to selected regular education classrooms, providing instructional support. Nicole is identified with socio-emotional disability and receives special education services for both language arts and mathematics. Kevin is diagnosed with autism, exhibits difficulties in social skills, and is cognitively high functioning. Hillary and Brady have specific

Students With Disabilities and the *Next Generation Science Standards*

learning disabilities and receive special education services for both language arts and mathematics. Jeff is also identified with specific learning disabilities and receives services for language arts. His math skills are advanced for his grade level. All of these students are part of the diverse community of learners working toward three-dimensional scientific understanding of the Earth-Sun-Moon relationship, as described in this vignette.

Exploring the Earth-Sun-Moon Relationship

Mr. O. initiated the unit by asking students to open their notebooks, write the numbers 1–8 down the next blank page, and title the page "Relative Diameters?" On the interactive whiteboard, he projected a slide from a multimedia presentation *Two Astronomy Games* that showed nine images, each identified by a letter and a label (Morrow 2004). The images were the Sun, Earth, a space shuttle, the Moon, the solar system, Mars, a galaxy, and Jupiter. Students were asked to number the objects in order from smallest (number 1) to largest (number 8) and from nearest to the surface of the Earth to farthest from the surface of the Earth. He planned to have students come back to this page later. Kevin seemed pleased and announced, "I love to study space!"

With a standard-size playground ball in hand, Mr. O. asked the class to imagine the ball was Earth and he wrote down the class' consensus of the ball's dimensions that they had figured out in math class. Then he presented the class with a box of seven balls in a variety of sizes and listed their dimensions on the interactive whiteboard. He asked: "If Earth was the size of this playground ball, which of these balls would be the size of the Moon?" One student (from each table) came up and chose the ball they thought would be correct. Their choices varied from a softball to a small marble. Before going further, the class reviewed the term *diameter* and Mr. O. asked, "If you know that Earth's diameter is 12,756 kilometers and the Moon's diameter is 3,476 kilometers, with your table groups, come up with a method to see if the ball you chose is the right size for this size Earth [holding up the playground ball]" (practice: Using Mathematics and Computational Thinking) (CC: Scale, Proportion, and Quantity).

After some discussion time, students reported their calculations. One group noticed that there was a proportional relationship in the diameters of approximately 1:4, Earth to Moon. A student asked how they made that determination. Jeff responded, "If you estimate using 12,000 and 3,000, three goes into twelve four times." He showed on the interactive whiteboard how four circles of the Moon fit across the diameter of the Earth. Mr. O. said, "Now think of your ball as a representation of the Moon and decide if you think it is the correct size. What can you do to be sure? Decide on a process." He let them use the playground ball as needed (DCI: MS-ESS1.A Earth's Place in the Universe).

Each group reported their findings and methods for determining whether or not their choice would be correct. One group made lines on paper where the endpoint of their ball was and did the same for the playground ball. Using those measurements and the 1:4 ratio, they decided if their Moon was the correct size. Another group used string to measure the diameter of the balls and then determined whether or not it was correct. Still another group held their ball up against the playground ball and moved their ball four times while marking the playground ball with a finger to see if their ball was the correct size for the model of Earth.

The groups reported their findings. Kevin was agitated as he explained, "I told my group they were not right. The racquetball is the only one that is possible as the Moon, but they wouldn't believe me." Mr. O. asked Kevin to restate the rule for when his group disagrees. Kevin thought and said, "When my group disagrees, I listen and then tell them what I think."

Only those groups with the racquetball had the correct size for the playground ball. Two of the students from one of those tables came up and showed how far they thought the Moon would be from Earth using the playground ball and racquetball model. Several students disagreed with the distance shown by the students. Four students came to the front, one by one, and showed their ideas about the distance between Earth and the Moon. Then Mr. O. showed them the actual distance from Earth to the Moon and the circumference of Earth in kilometers. He asked them again to use the new evidence to determine how to figure out the distance in the model and to show it using string. Students were shocked at the distance the Moon was from Earth in this model. Their estimates had been much lower.

As the class finished presenting their arguments for the correct size balls for the Sun and Earth, students considered the relative size of the Sun and the distance of the Sun from Earth in the model. They used the evidence of the diameter of the Sun and its distance from Earth in the same way they determined the size and distance of the Moon from Earth. Some students were surprised at the size of the Sun and its distance from Earth in this model. Jeff decided that they could not fit the Sun in the room. He explained that it would take over 100 playground balls to approximate the Sun's diameter. Jeff was eager to share his mathematical skill at finding the answer: "I know the answer! It would take almost 12,000 playground balls lined up to show how far away the Sun would be in this model." Two students nonchalantly said, "That's a lot," and "The Sun is very far away from Earth" (CC: Scale, Proportion, and Quantity).

The students returned to their initial ideas on the "Relative Diameters" page in their notebooks, renumbered the objects, and recorded any ideas that had changed after making the model. After giving students time to write their responses, Mr. O. showed images of the items on the interactive whiteboard and led a discussion of the great distances between objects in the solar system in preparation for modeling the Moon's phases (DCI: MS-ESS1.B Earth's Place in the Universe).

For this lesson sequence, Mr. O. considered the makeup of the table groupings of students. He wanted the special education and other struggling students to have support while determining methods to check their choice of the Moon model, so he grouped students with that concern in mind. He used physical representations of Earth and the Moon and had students represent the distance physically, thereby assisting them in visualization

Students With Disabilities and the *Next Generation Science Standards*

and comprehension. *(The strategy of providing multiple means of representation was important to support understanding for his special education students, but it also benefited all of his students.)*

Exploring Moon Phases

Mr. O. showed how the Moon's and Earth's orbital planes are offset by 5 degrees in an effort to help students understand how light can illuminate the Moon when it is on the other side of Earth without being blocked by Earth's shadow. Throughout this instruction the special education students were strategically placed at tables in groups that would support their engagement in the content and activity.

Mr. O. downloaded an open-source planetarium program onto his whiteboard-connected computer as well as onto the 14 student computers he had in his classroom. On the first day of Moon phase instruction, each student received a one-page Moon calendar similar to one they took home. The students who had completed the calendar kept it out to compare their observations to the data collected using the software. Mr. O. launched the program on the interactive whiteboard, introduced the students to the software, and showed them how to change the date and set up the scale Moon so they could see the phases.

Recording began on the first Sunday on the calendar and ended on the last Saturday, resulting in five weeks worth of data to analyze (practice: Analyzing and Interpreting Data). Mr. O. modeled how to record the data on the whiteboard next to the interactive whiteboard. Students recorded the time and location of moonrise and moonset as well as the apparent shape of the Moon in the sky for each date. To make sure that students understood the process and were recording accurately, he walked through the room and checked student work throughout the lesson. Also during this modeling process, the students paid attention to the Sun-Moon relationship so they could see the light from the Sun traveling in a straight line to the Moon. The Moon was in the sky as the Sun was rising, and they focused on the Moon so that they could use the model for predictions. Mr. O. asked, "Does anyone know where the Sun is right now?" Brady responded, "It's more to the east and still rising." Using the time and date function in the program, he advanced the time to show the sunrise and said, "Look at the Sun and Moon. What pattern do you notice about the light on the Moon in relation to the Sun?" (CC: Patterns). Hillary answered, "It is going from the Sun to the Moon." Mr. O. responded, "Hmm. The light travels in a straight path from the Sun to the Moon. You have already learned that light travels in a straight line. Can we use that information to predict the position of the Sun even if we can't see it? Let's try as we continue."

After a few days' worth of data were collected, Mr. O. asked students to predict the time and direction for moonrise and moonset and brought their attention to the patterns in the data. He asked, "What time do you think the Moon will set on this day? The last time was 12:09." Mark said, "I think 12:59." Mr. O. advanced the time until the Moon set—at 13:08. Jeff called out, "So it is setting about an hour later each time." A student said, "So let's see

CHAPTER 8

if that pattern continues the whole month." Once the students had a foundation for data collection (about 8–10 days), they went to the computers in partners so they could work more independently to complete the data collection on the calendar.

Mr. O. wanted some control over the assignment of partners to provide support for students who needed it and to challenge more advanced students, so he predetermined the partners and assigned them before sending them to the computers. Jeanette and Nicole worked with their paraprofessional. As a modification to recording the data, they were given a calendar with a set of Moon phase images. As they worked with the paraprofessional, Jeanette said, "When do we write the answer?" Nicole answered, "You have to wait and look at Stellarium and glue the picture." The paraprofessional redirected Nicole and made sure that the directions were understood: Match the image to the one on Stellarium and glue it on the calendar for each day. They did not record the moonrise and moonset times. Hillary, Jeff, and Brady were each paired with a partner whose academic abilities were a little higher than their own, allowing them to receive some support from the partner. Kevin was paired with someone at the same ability level who would be patient with his unique social skills. Kevin enthusiastically stated, "I love science and I love to learn about space."

While students worked at the computers to complete the calendar, Mr. O. took aside small groups of students to do an activity in which they modeled Moon phases using Styrofoam balls, their heads, and a lamp with a bare bulb. Students stood in a circle around the lamp representing the Sun, holding a Styrofoam ball on a stick representing the Moon. They held the ball at arm's length and rotated their bodies using their heads as a representation of Earth so they could see the Earth view of the Moon in all its phases in the lit portion of the ball. The students went through the phases, naming each one and making sure that all students could see the lit portion on the Styrofoam balls for each phase.

Jeanette kept turning the wrong way as she looked at the student across from her. "Is this the way?" she asked, as Mr. O. gently helped direct her turn. Nicole was focused on the computer groups, so Mr. O. directed Nicole to look at the Styrofoam ball and the changing shadow. "What? I don't see the shadow," she said. Mr. O. pointed out the curve of light on the Moon. "I see it!" Nicole said.

Small groups allowed Mr. O to make sure that all students were able to accurately illustrate the phases in the model, giving him the opportunity to physically move them into position as necessary. In addition, he kept students from the first group who he felt might need more time with the model in the second group for more practice if needed.

The students collaborated to explain how the model of the Moon phases illustrated changes in the apparent shape of the Moon. They discussed limitations of the models—the things that a model is unable to show accurately. The students identified the relative sizes of the Sun, Earth, and Moon as well as the relative distances between each as being inaccurate in this model (practice: Developing and Using Models).

Students With Disabilities and the *Next Generation Science Standards*

To finish the class period, all students were at the computers working with Stellarium and their calendars. Mr. O. walked around the room assisting students with their data collection. Jeanette called Mr. O. over and quietly said, "I lost the Moon and can't find it." He showed her how to search for it using the "find" function. Many of the students had changed the dates, so he stopped the class to note, "Many of you have found that this program shows future dates." To reinforce the language Mr. O. had used on many occasions throughout the unit, he asked, "What does that tell us about the planets and the Moon? They all move …" and students responded, "… in predictable patterns."

Over the next two days while students continued working on their calendar with Stellarium, Mr. O. again pulled small groups of students to use another model showing Moon phases (practice: Developing and Using Models). This one used golf balls that were painted black on half of the sphere, leaving the other half showing the side of the Moon lit by the Sun (Young and Guy 2008). The golf balls were drilled and mounted on tees so they would stand up on a surface. Mr. O. had two sets—one set up on a table that showed the Moon in orbit around the Earth in eight phase positions as the "space view" model (Figure 8.1), and the other with the model Moons set on eight chairs circled in the eight phase positions to show the "Earth view" model

FIGURE 8.1.

SPACE VIEW MODEL

First, students were shown the space view model and asked what they noticed about the Moons. Mr. O. wanted them to notice that the white sides of all the balls (showing light) faced the same direction. He asked them to identify the direction of the Sun. Nicole was looking toward the window, and Mr. O. asked her, "Nicole, where is the Sun in our model here in the classroom?" Nicole looked around and responded, "Over here, I think, because that's where the lit up sides are facing."

Then Mr. O. drew the students' attention to the model on the chairs, the Earth view model. All the balls in this model faced the same direction as those in the space view model. Students again identified the direction of the Sun and noted that the position of the Moons in both models was the same (DCI. MS-ESS1.A Earth's Place in the Universe). One at a time, students physically got into the center of the circle of chairs and viewed the phases at eye level (Figure 8.2, p. 90), which simulated the Earth view of each phase. (*Providing multiple means of action and expression is one of three principles of Universal Design for Learning.*)

CHAPTER 8

FIGURE 8.2.

THE EARTH VIEW MODEL

Each of the students with Individualized Education Plans (IEPs) was put in a different small group, with the exception of Jeanette and Nicole, who were in the same small group. Their turn inside the circle was last, giving them the opportunity to observe, listen, and practice while verbalizing the phases and location of the Sun within the system. This activity made the diagram, often found in books and worksheets showing both views on the same diagram, less confusing to the students.

Although most students were not finished with the calendar, Mr. O. brought all students together the next day to create a foldable showing the Earth view of the Moon phases similar to diagrams found in books. Students created their Moon phases using eight black circles and four white circles, cutting the white circles to make two crescent Moons, two gibbous Moons, and two quarter Moons. The white circle pieces were placed on the black circles to create the phases, and later glued on the foldable. Jeanette was unsure of the placement of the pieces. "Where does this one go?" Jeanette asked referring to the gibbous Moons, which were incorrectly placed. "Look at mine. I'm right," said Nicole who also had confused the two phases. As he walked around the room checking student work, Mr. O. gently pointed out the lit side of the Moon and asked which phase that represented. Inside the foldable, students drew a large circle to represent the Moon. (*Providing multiple means of representation is one of the three principles of Universal Design for Learning.*)

They partnered to read *The Moon* by Seymour Simon (2003). Students used the information in the book to label the Moon phases on their foldable, write about the Moon's surface, and record any new questions that arose from their reading. Kevin asked, "When is the next solar and lunar eclipse?" Jeanette questioned, "What samples were brought back from the Moon?" And Nicole wanted to know, "Where did Americans land on the Moon?"

To support their reading of the text, Hillary, Brady, and Jeff were given the option of being paired with students who had more advanced reading skills or using Mr. O.'s recordings made on handheld computers. Jeanette and Nicole had the support of their paraprofessional in reading and obtaining information from the text. Mr. O. asked Kevin, "What would you prefer?" He answered, "Oh, I think this time I want to read by myself because I love space and want to find out more about the Moon." As students finished their reading and writing, they went back to finish their calendars using the software.

Students finished the calendar at different rates. When finished, they checked their work against the calendar that Mr. O. had completed. Since several pairs finished at the same time, he grouped the pairs to discuss the patterns they noticed in their calendars. He gave

Students With Disabilities and the *Next Generation Science Standards*

them a list of questions to guide their discussion and asked them to conference with him when they were finished. (*Providing multiple means of engagement is one of the three principles of Universal Design for Learning.*) He expected all students to observe that the lit segment of the Moon's face increased, decreased, and increased again relative to the part in shadow. He also expected students to notice that the lit side of the Moon was on the left after the full Moon phase and on the right after the new Moon phase, as viewed from Earth. Students who finished with all tasks were allowed to use text materials and internet resources to research answers to the questions they developed when reading *The Moon*, while the rest of the students completed their calendars.

Assessing Student Learning

Throughout the lesson sequence, Mr. O. continually assessed students' progression through observations and conferences. If he noticed students needed more experience with Moon phases, he provided them with additional activities such as videos and Moon phase cards. In one formal assessment of understanding, Mr. O. paired students together so that one was assigned to be the Earth and the other the Moon. He designated one wall of the classroom as the Sun and then asked the Moons to show different phases. The students switched roles so that Mr. O. could assess everyone. He also used this model to demonstrate the Moon's coincident rotation and revolution. In another formal assessment, he asked students to draw a model on whiteboards showing the relationship of the Earth, Moon, and Sun in full Moon phase.

NGSS CONNECTIONS

NGSS require that students engage in science and engineering practices to develop deeper understanding of the disciplinary core ideas and crosscutting concepts. This presents both challenges and opportunities to special education students, since a broad range of disabilities impacts their science learning. This vignette highlights examples of strategies that support all students while engaging in science practices and in rigorous content. The lessons give students varied exposure to the core ideas in space science, helping to prepare all students to demonstrate mastery of the three components described in the *NGSS* performance expectation. See Figure 8.3 (p. 95) for the comprehensive list of *NGSS* and *CCSS* from the vignette.

CHAPTER 8

Performance Expectations

MS-ESS1-1 Earth's Place in the Universe

Develop and use a model of the Earth-Sun-Moon system to predict and describe the cyclic patterns of lunar phases, eclipses of the Sun and Moon, and seasons.

MS-ESS1-3 Earth's Place in the Universe

Analyze and interpret data to determine scale properties of objects in the solar system.

Disciplinary Core Ideas

ESS1.A The Universe and Its Stars

Patterns of the apparent motion of the Sun, the Moon, and stars in the sky can be observed, described, predicted, and explained with models.

ESS1.B Earth and the Solar System

The solar system consists of the Sun and a collection of objects, including planets, their Moons, and asteroids that are held in orbit around the Sun by its gravitational pull on them.

Science and Engineering Practices

Developing and Using Models

Develop and use a model to describe phenomena.

Analyzing and Interpreting Data

Analyze and interpret data to determine similarities and differences in findings.

Students With Disabilities and the *Next Generation Science Standards*

Students were engaged in a number of science practices with a focus on Developing and Using Models and Analyzing and Interpreting Data. Space science lends itself well to the use of models to describe patterns in phenomena and to construct explanations based on evidence. With guidance from their teacher, students used the ratios of the diameters of Earth and its Moon to construct a class model of the relative sizes of the two objects. Using distance and Earth's diameter or circumference ratios, they also constructed a distance model of those objects. In addition, the relative size of the Sun and the relative distance from Earth in this model was calculated and described, although not constructed (due to the constraints of the room and location). Throughout the vignette, a variety of models were used to help students identify patterns in the relative positions of the Earth, Moon, and Sun, and to explain Moon phases.

Crosscutting Concepts

Patterns
Patterns can be used to identify cause-and-effect relationships.
Scale, Proportion, and Quantity
Time, space, and energy phenomena can be observed at various scales using models to study systems that are too large or too small.

Students made predictions about the data collected and recorded them on the calendar, using the lens of the crosscutting concept of Patterns. When analyzing and interpreting the data, they identified the patterns in the Earth-Sun-Moon relationship. The pattern made by the lit portion of the Moon was observed and recorded. In addition, students considered the crosscutting concept of Scale, Proportion, and Quantity as they constructed models of relative sizes and distance of the Sun and planets.

CCSS CONNECTIONS TO ENGLISH LANGUAGE ARTS AND MATHEMATICS

Students used the text in *The Moon* (Simon 2003) to label each phase of the Moon and summarize information about the surface of the Moon in their graphic organizer foldable. This reading and writing connects to the *CCSS ELA*:

- **RST.6-8.1** *Cite specific textual evidence to support analysis of science and technical texts.*
- **WHST.6-8.2** *Write informative/explanatory texts to examine a topic and convey ideas, concepts, and information through the selection, organization, and analysis of relevant content.*

CHAPTER 8

When comparing sizes and distances, students were challenged to find ways of comparing numbers, applying the *CCSS Mathematics* MP.1. In addition, students used rounding and estimation to calculate the quotients in the ratios, both skills developed in earlier grades and used again in fifth grade, standard 4.OA. Throughout the unit, students reasoned quantitatively as they compared the sizes of the Earth and Moon, standard MP.2. As students made conclusions about which ball was the Moon, they argued for their selection and agreed or disagreed with each other using their calculation, standard MP.3:

- **6.RP.A.1** *Understand the concept of a ratio and use ratio language to describe a ratio relationship between two quantities.*
- **MP.1** *Make sense of problems and persevere in solving them.*
- **MP.2** *Reason abstractly and quantitatively.*
- **MP.3** *Construct viable arguments and critique the reasoning of others.*

EFFECTIVE STRATEGIES FROM RESEARCH LITERATURE

Students with disabilities have IEPs, specific to the individuals, that mandate the accommodations and modifications that teachers must provide to support their learning in the regular education classroom. By definition, accommodations allow students to overcome or work around their disabilities with the same performance expectations of their peers, whereas modifications generally change the curriculum or performance expectations for a specific student (National Dissemination Center for Children with Disabilities 2010). Special education teachers can be consulted to provide guidance for making accommodations and modifications to help students with IEPs succeed with the *NGSS*.

Two approaches of providing accommodations and modifications are widely used by general education teachers in their classrooms. *Differentiated instruction* is a model in which teachers plan flexible approaches to instruction in the following areas: content, process, product, affect, and learning environment (Institutes on Academic Diversity 2009–2012). This vignette highlights Universal Design for Learning as a framework with a set of principles for curriculum development that provides equal access to all learners in the classroom (CAST 2012). The framework supplies a set of guidelines for teachers to use in curriculum planning that is organized around three principles: (1) to provide multiple means of representation, (2) to present multiple means of action and expression, and (3) to encourage multiple means of engagement. Teachers identify barriers that their students may have to learning and then use the framework to provide flexible approaches of instruction. While both differentiated instruction and Universal Design for Learning benefit students with disabilities, they also benefit all students.

Students With Disabilities and the *Next Generation Science Standards*

FIGURE 8.3.

NGSS AND *CCSS* FROM VIGNETTE

MS-ESS1 Earth's Place in the Universe

Students who demonstrate understanding can:

MS-ESS1-1. Develop and use a model of the Earth-Sun-Moon system to predict and describe the cyclic patterns of lunar phases, eclipses of the Sun and Moon, and seasons.

MS-ESS1-3. Analyze and interpret data to determine scale properties of objects in the solar system.

The performance expectations above were developed using the following elements from the NRC document *A Framework for K–12 Science Education*:

SCIENCE AND ENGINEERING PRACTICES	DISCIPLINARY CORE IDEAS	CROSSCUTTING CONCEPTS
Developing and Using Models Modeling in 6–8 builds on K–5 and progresses to developing, using, and revising models to support explanations, describe, test, and predict more abstract phenomena and design systems. • Develop use a model to describe phenomena. **Analyzing and Interpreting Data** Analyzing data in 6–8 builds on K–5 experiences and progresses to extending quantitative analysis to investigations, distinguishing between correlation and causation, and basic statistical techniques of data and error analysis. • Analyze and interpret data to determine similarities and differences in findings.	**ESS1.A: The Universe and Its Stars** • Patterns of the apparent motion of the Sun, the Moon, and stars in the sky can be observed, described, predicted, and explained with models. **ESS1.B: Earth and the Solar System** • The solar system consists of the Sun and a collection of objects, including planets, their moons, and asteroids that are held in orbit around the Sun by its gravitational pull on them.	**Patterns** • Patterns can be used to identify cause-and-effect relationships. **Scale, Proportion, and Quantity** • Time, space, and energy phenomena can be observed at various scales using models to study systems that are too large or too small.

CCSS Connections for English Language Arts and Mathematics

RST.6-8.1 Cite specific textual evidence to support analysis of science and technical texts.

WHST.6-8.2 Write informative/explanatory texts to examine a topic and convey ideas, concepts, and information through the selection, organization, and analysis of relevant content.

6.RP.A.1 Understand the concept of a ratio and use ratio language to describe a ratio relationship between two quantities.

MP.1 Make sense of problems and persevere in solving them.

MP.2 Reason abstractly and quantitatively.

MP.3 Construct viable arguments and critique the reasoning of others.

CHAPTER 8

CONTEXT

DEMOGRAPHICS

The number of children and youth age 3–21 receiving special education services under the Individuals with Disabilities Education Act (IDEA) rose from 4.1 million in 1980 (10% of student enrollment) to 6.7 million in 2005 (14% of student enrollment) (National Center for Education Statistics 2011). By 2009, that number had decreased to 6.5 million (13% of student enrollment). Special education services under IDEA are provided for eligible children and youth who are identified by a team of professionals as having a disability that adversely affects academic performance.

Students with disabilities are also protected under Section 504 of the Rehabilitation Act of 1973, which covers all persons with a disability from discrimination in educational settings based solely on their disability. Section 504 requires a documented plan in which a school provides reasonable accommodations, modifications, supports, and auxiliary aides to enable students to participate in the general curriculum, although it does not require students to have an IEP. Since the implementation of Public Law 94-142 enacted in 1975, there has been concern about disproportionate representation of racial and ethnic minorities, economically disadvantaged students, and English language learners in special education programs (Donovan 2002; U.S. Commission on Civil Rights 2009). While there continues to be a disproportionate number (both overrepresentation and underrepresentation) of different populations of students identified in special education within general and specific disability categories, determining the factors that affect this inequality is difficult and complex.

SCIENCE ACHIEVEMENT

On the National Assessment of Educational Progress (NAEP) in science, the gap in grade 12 scores between students with disabilities and students with no disabilities has persisted at 38 points in 1996, 39 points in 2000, and 37 points in 2005. The grade 8 gap has continually decreased from 38 points in 1996, to 34 points in 2000, and to 32 points in 2005. The grade 4 gap increased from 24 points in 1996 to 29 points in 2000 before it finally decreased to 20 points in 2005. The results indicate two important points. First, while achievement gaps persisted across the three grade levels, patterns of increase or decrease were inconsistent at each grade level. Second, achievement gaps were wider as students advanced to higher grade levels.

In 2009, the NAEP science achievement gaps between students with disabilities (including those with 504 plans) and students with no disabilities were 32 points at grade 12, 30 points at grade 8, and 24 points at grade 4. This confirms that achievement gaps were wider as students advanced to higher grade levels, consistent with results in 1996, 2000, and 2005 described above.

Students With Disabilities and the *Next Generation Science Standards*

The NAEP did not allow accommodations for students with disabilities prior to 1996. In 1996, some schools were allowed to use accommodations for students with disabilities while others were not allowed to assess the impact on NAEP results. In a continuing effort to be more inclusive, guidelines were developed that specified that students with disabilities should be included in the NAEP assessment. Despite attempts to standardize the inclusion process, exclusion rates vary across states (Stancavage, Makris, and Rice 2007).

Thus, all students with disabilities are not included in the NAEP science assessment, making it difficult to identify accurate achievement gaps between students with disabilities and their peers. In addition, the data are not disaggregated according to disability category, further complicating the process to identify spe cific achievement gaps. The National Assessment Governing Board recommended that NAEP should report separately on students with IEPs and those with 504 plans and should count only students with IEPs as students with disabilities. Prior to 2009, NAEP's "students with disabilities" category included both students with IEPs and students with 504 plans. In 2009, although students with 504 plans received accommodations according to their plans, their scores were reported in the category of students without disabilities.

EDUCATION POLICY

Enacted in 1975, Public Law 94-142, Education for All Handicapped Children Act, mandated the provision of a free and appropriate public school education in the least restrictive environment for children and youth ages 3–21 with disabilities. Public schools were required to develop an IEP with parental input that would be as close as possible to a non-handicapped student's educational experience. The IEP specifies the types and frequencies of services to be provided to the student, including speech-language; psychological, physical and occupational therapy; and counseling services. It specifies the accommodations and modifications that are to be provided for the student in curriculum, instruction, and assessment. The IEP also described the student's present levels of academic performance and the impact of disabilities on performance.

Students with disabilities are also protected under Section 504 of the Rehabilitation Act of 1973. While special education services under IDEA [IDEA is described in more detail in the following paragraph] are provided for eligible children and youth who are identified by a team of professionals as having a disability that adversely affects academic performance, Section 504 covers all persons with a disability from discrimination in educational settings based solely on their disability. Section 504 does not require an IEP, but does require a documented plan in which the school provides reasonable accommodations, modifications, supports, and auxiliary aides to enable the student to participate within the general curriculum.

In 1990, Public Law 94-142 was revised and renamed Individuals with Disabilities Education Act (IDEA). The most recent revision and reauthorization was completed in 2004

CHAPTER 8

with implementation in 2006. One notable change is the requirement that state-adopted criteria to identify students who have Specific Learning Disabilities (SLD) must not require a severe discrepancy between intellectual ability and achievement; must permit the use of a process based on the child's response to scientific, research-based intervention; and may permit the use of other alternative research-based procedures.

SLD, as a category, has the largest number of identified students and is defined by IDEA in the following way:

> *The term "specific learning disability" means a disorder in one or more of the basic psychological processes involved in understanding or in using language, spoken or written, which disorder may manifest itself in the imperfect ability to listen, think, speak, read, write, spell, or do mathematical calculations ... Such term includes such conditions as perceptual disabilities, brain injury, minimal brain dysfunction, dyslexia, and developmental aphasia ... Such term does not include a learning problem that is primarily the result of visual, hearing, or motor disabilities, of mental retardation, of emotional disturbance, or of environmental, cultural, or economic disadvantage.* (TITLE I/A/602/30)

Under Elementary and Secondary Education Act regulations (ESEA 1965), students with disabilities are monitored for Adequate Yearly Progress (AYP) in the content areas of language arts and mathematics, with increased accountability expected as special education services continue (ESEA Title 1, Part A, Subpart 1. Sect 1111.b.2.C.V.II.cc.). Data on students' science progress are also collected and reported once at the elementary school level, middle school level, and high school level. In 2007, final regulations under ESEA and IDEA were released to allow more flexibility to states in measuring the achievement of students with disabilities (34 C.F.R. Part 200; U.S. Department of Education 2007).

The U.S. Office of Special Education created the IDEA Partnership to promote collaboration among the many national and state agencies and stakeholders dedicated to improving outcomes for students with disabilities. In response to the growing concern about increasing numbers of students identified with learning disabilities, there has been a call for identifying students at risk and implementing scientific, research-based intervention. The response to intervention (RTI) model is an effort to improve early intervention for students while improving learning outcomes and reducing the number of students identified as learning disabled.

REFERENCES

Center for Applied Special Technology (CAST). 2012. Universal Design for Learning. Wakefield, MA. *www.udlcenter.org*

Students With Disabilities and the *Next Generation Science Standards*

Center on Response to Intervention (RTI) at American Institutes for Research. Washington, DC. *http://state.rti4success.org/index.php*

Donovan, S. 2002. *Minority students in special and gifted education.* Washington, DC: National Academies Press.

Elementary and Secondary Education Act of 1965, Pub. L. No. 89–10, 79 Stat. 27.

IDEA Partnership. *www.ideapartnership.org/index.php?option=com_content&view=category&layout=blog&id=15&Itemid=56*

Institutes on Academic Diversity. 2009–2012. Differentiation Central. *www.diffcentral.com/index.html.*

Morrow, C. 2004. Two astronomy games. Space Science Institute science education website. *www.spacescience.org/education/instructional_materials.html*

National Assessment Governing Board. 2010. *NAEP testing and reporting on students with disabilities and English language learners. www.nagb.org/content/nagb/assets/documents/policies/naep_testandreport_studentswithdisabilities.pdf*

National Center for Education Statistics (NCES). 2011. *The condition of education 2011* (NCES 2011-033). Washington, DC: U.S. Department of Education. *http://nces.ed.gov/pubs2011/2011033.pdf.*

National Dissemination Center for Children with Disabilities. 2010. Supports, modifications, and accommodations for students. *www.parentcenterhub.org/repository/accommodations*

Simon, S. 2003. *The Moon.* New York: Simon and Schuster.

Stancavage, F., F. Makris, and M. Rice. 2007. *SD/LEP inclusions/exclusion in NAEP: An investigation of factors affecting SD/LEP inclusions/exclusions in NAEP.* Washington, DC: American Institutes for Research.

U.S. Commission on Civil Rights. 2009. *Minorities in special education.* Washington, DC: U.S. Commission on Civil Rights. *www.usccr.gov/pubs/MinoritiesinSpecialEducation.pdf*

U.S. Department of Education. 2007. *Modified academic achievement standards.* Washington, DC: U.S. Department of Education. *www2.ed.gov/policy/speced/guid/modachieve-summary.html*

Young, T., and M. Guy. 2008. The Moon's phases and the self shadow. *Science and Children* 46 (1): 30.

Zike, D. DMA: Dinah-might adventures. *www.dinah.com*

CHAPTER 9

ENGLISH LANGUAGE LEARNERS AND THE *NEXT GENERATION SCIENCE STANDARDS*

MEMBERS OF THE *NGSS* DIVERSITY AND EQUITY TEAM

ABSTRACT

The number of English language learners in schools across the nation has increased dramatically over the past decade. At the same time the gap in science proficiency between English language learners and non-English language learners has widened. As a result of Elementary and Secondary Education Act (ESEA) legislation, school districts are held more accountable for English language learners' progress and must seek ways to provide equal access to education. The *Next Generation Science Standards* (*NGSS*) set high expectations in science for all students, and teachers of English language learners must employ effective strategies to deepen understanding of science while learning English. The literature indicates five areas where teachers can support science and language for English language learners: (1) literacy strategies for all students, (2) language support strategies with English language learners, (3) discourse strategies with English language learners, (4) home language support, and (5) home culture connections. The vignette highlights how these strategies promote English language learners' understanding of disciplinary core ideas, science and engineering practices, and crosscutting concepts as described by the *NGSS*.

VIGNETTE: DEVELOPING AND USING MODELS TO REPRESENT EARTH'S SURFACE SYSTEMS

While the vignette presents real classroom experiences of the *NGSS* implementation with diverse student groups, the following considerations should be kept in mind. First, for the purpose of illustration only, the vignette is focused on a limited number of performance expectations. It should not be viewed as showing all instruction necessary to prepare students to fully understand these performance expectations. Neither does it indicate that the performance expectations should be taught one at a time. Second, science instruction should take into account that student understanding builds over time and that some topics or ideas require extended revisiting through the course of a year. Performance expectations will be realized by using coherent connections among disciplinary core ideas, science

CHAPTER 9

and engineering practices, and crosscutting concepts within the *NGSS*. Finally, the vignette is intended to illustrate specific contexts. It is not meant to imply that students fit solely into one demographic subgroup, but rather it is intended to illustrate practical strategies to engage all students in the *NGSS*.

INTRODUCTION

The science and engineering practices described in the *NGSS* are language intensive and present both language demands and language learning opportunities for English language learners. This vignette illustrates how very young students, many of whom are English language learners, can develop proficiency in these language-intensive science practices while engaging with rigorous science content. With a focus on Earth science, the teacher powerfully demonstrates how her students overcome language barriers to use models, develop claims, and explain their reasoning using evidence. The vignette highlights effective strategies to provide English language learners' access to core ideas, practices, and crosscutting concepts of science. Throughout the vignette, classroom strategies that are effective for all students, particularly for English language learners according to the research literature, are highlighted in parentheses.

ELL Connections

Like all of the classes at Monroe Elementary, a school with more than 74% of the population at or below the poverty level, Ms. H.'s second-grade class was made up of diverse groups of learners. Her class included three Hmong students, eight African Americans, three students who recently arrived from Gambia, two from Mexico, and two Mexican Americans. Of her 18 children, 9 were English language learners.

FIGURE 9.1.

STUDENT WRITING CLAIMS ON THE WHITEBOARD

Three weeks into the Earth science unit, Ms. H. introduced a task in which students had to rely on their team members, their field notes, and their diagrams of soil profiles to match three different types of soil to their source locations. Each team of three students had paper plates piled with soil on the table in front of them and three location cards. The unidentified soil samples came from sites within walking distance of Monroe School. One card was labeled "Urban Marsh" and had a picture of the marsh near the school. The other cards were labeled "Coniferous Hill" and "School Yard Field" with respective photos. (*Ms. H. used these labeled cards with photos to represent concepts as a language support strategy for English language learners.*)

English Language Learners and the *Next Generation Science Standards*

Throughout the unit, each team had dug small soil pits in the three locations, recorded data about soil composition and function (e.g., infiltration), and designed, to scale, diagrams of soil profiles for each site (practices: Analyzing and Interpreting Data, and Using Mathematics and Computational Thinking). Now, with their teams, they were making claims, looking for evidence to support their claims, and recording their reasoning on large white boards (Figure 9.1; practice: Engaging in Argument From Evidence).

One group—Kaleem, Moustafa, and Victor—was working closely together. Kaleem consulted his partners, who were sifting through one of the soil samples and looking for clues. "Feel this!" Kaleem said. "Feel this, you guys; this is so sandy. Here, feel." He handed one of the plates of soil to Victor and asked, "Don't you think that is sandy?" He looked at the open page in his science notebook for his notes (Figure 9.2). The class had previously tested and written about the compositions of different soils.

Victor and Moustafa were focused on something they had discovered in one of the soils: pine needles. Moustafa, a recent newcomer from Gambia, said, "This is, this thing, this is …" He held up the pine needles. Kaleem supplied the term *pine needle*. Then Kaleem added dramatically, "Oh! Coniferous hill."

The boys tramped over to the wall sized diagrams that the class had made of the soil profiles for each location (Figure 9.3). These diagrams served as models to help the students think about and explain differences in soil. Only one of the soil profiles showed pine needles in the soil, and that was the coniferous hill. Victor looked at his coniferous hill notes and said, "There are small roots three [inches] down. And pine needles one [inch] down." Victor used the team's field notes to reinforce his thinking about the type of soil.

FIGURE 9.2.
FIELD NOTES

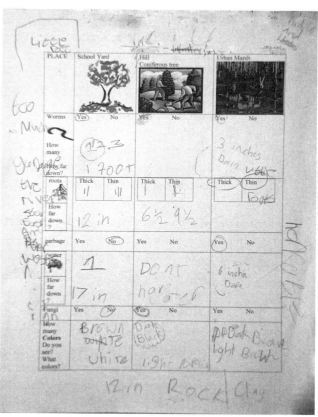

FIGURE 9.3.
SOIL PROFILE MODEL

CHAPTER 9

Moustafa pointed to the colors on another adjacent model and said, "And the soil was dark, dark, dark black, and it looks like black." Kaleem responded to Moustafa, "What do you mean?" Moustafa said, pointing to the soil and running his finger along the soil profile for the urban marsh, "Here, urban marsh, it is wet soil, dark, dark black." He looked for the label "organic" on the diagram, found it, and added, "Organic soil." Moustafa's specialization was the urban marsh; his group members knew this and nodded appreciatively. Focusing on the colors of the soil on the paper plates, they looked for the darkest soil from the urban marsh. They were well on their way to solving the mystery (practice: Developing and Using Models).

Each member of the three-person teams had developed expertise in a different soil site. Earlier, Moustafa and his fellow "urban marsh specialists" had studied the marsh and its soil, and refined the class model of the urban marsh. The urban marsh specialists had read books from the library, written and drawn about their site in their science notebooks, and presented their findings to the class (practice: Obtaining, Evaluating, and Communicating Information). Now Moustafa was working with Victor and Kaleem, experts from each of the other sites, on the problem of matching the soils to their locations. The team depended on each member's knowledge to aggregate all the evidence the group needed to support their claims. Ms. H. often used the jigsaw format to fuel contributions of her English language learners. (*The jigsaw is an example of an effective discourse strategy for English language learners.*) Since every student possessed unique and vital information, the format validated each student's contribution.

With Kaleem and Moustafa's help, Victor, the "coniferous hill specialist," used the sentence frame: "This soil came from _site_ because _evidence_ " to write his explanation n the white board under plate #2. As the students had taken care to clearly and correctly label all of the elements on the model, Victor double-checked his spelling against the model labels: "*This soil came from the coniferous hill because it shows there is pine needles at that place and there is pine needles in the dirt.*" Victor and his team needed to come up with two more pieces of evidence to support this claim (practice: Engaging in Argument From Evidence).

How did Ms. H. help her students become so interested in soil? Ms. H. began her unit three weeks previously by soliciting prior knowledge. She had the students collect a small sample of soil from the school yard field in paper cups, bring it into the classroom, and describe the soil to their partners. Ms. H. recorded their observations in a conceptual web to organize thinking and build schema about the topic. (*The conceptual web is an effective literacy development strategy for English language learners.*) She had four categories for the web: where soil is; what soil looks, feels, and smells like; what soil is made of; and experiences with soil. Students took turns passing the squishy ball around the circle, sharing ideas and questions about soil, or choosing to pass. If a student asked a question, Ms. H. wrote the question on a sentence strip and taped it on the cabinet under the heading, "Scientific Questions About Soil."

English Language Learners and the *Next Generation Science Standards*

She was delighted to see that each student had at least one question or idea to share. Moustafa said that the soil was black, brown, and white. Some students shared experiences with soil in a garden, and others talked about worms. Julissa asked, "How does soil grow?" And Edrissa asked, "Does soil come from rocks?" When the list was exhausted, each student chose one idea to draw and illustrate in their science notebooks; some students had time to write down two or three.

The next day Ms. H. taped a sentence strip after the words "Scientific Question" on the whiteboard: "Is soil the <u>same</u> everywhere?" The class used "turn and talk" to discuss the question with their partners. Ms. H. made sure that students with similar language backgrounds had the opportunity to discuss the question in their home language or English. (*English language learners benefit from home language support in the classroom.*) Each student's goal was listening to the partner. Ms. H. asked, "Did everyone get a chance to share?" and then added, "You have one more minute." Two students needed to switch listener and speaker roles. She picked three craft sticks with the students' names on them, and placed the names on the carpet for all to read. She asked them, "What did your partner say?"

Trinique's name was picked first and she said, "My partner Deshawna said that soil is not the same everywhere because it feels different at the beach." Ms. H. checked with Deshawna, "Is that what you said, Deshawna?" who nodded and added, "It is rough, sand feels rough and scratchy." Ms. H. wrote under the scientific question, "Soil at the beach feels rough." After soliciting two more responses, Ms. H. asked students to draw and write about the scientific question: "Is soil the same everywhere?" in their science notebooks.

The conceptual web from the day before supported the students' writing as they used it to find vocabulary and check spelling. Jesus, an English language learner, wrote, "Soil are different place like Canada, New York, and Florida because there are different towns, city, or country." Zytasia wrote, "No, because there's hard soil and fresh soil, because I saw white soil and black soil."

The scientific question written in students' home language was assigned for homework so that they could talk about the question with their parents. Students were permitted to fill in the answer in their home language or English. Ms. H. often asked the students to conduct parent interviews because she wanted to spark discussions at home. (*English language learners benefit from home language support in the classroom.*) Ms. H. was able to construct new meaning with the resources her students brought from home. Many students had important thoughts and questions that they could discuss using the science-specific vocabulary in their home language.

When the students shared their parent interviews with their classmates, the answers varied from "no, soil comes in different bags," to an answer written in meticulous printing, "Gambia—the soil is red and dusty, some places have good soft clay and are good for farming." One Hmong boy talked to his family and asked the school's translator to write their answer in Hmong and English, "Not the same. Some soil is sticky (clay), some is dry.

CHAPTER 9

Some soil is black and some is yellow." Jesus's mom had helped him write her answer in Spanish, and he proudly read it to the class: "*No, no es igual porque en unos lugares la tierra es seca y dura y en otros lugares es húmeda.*" Based on the interviews, it was clear that the homework assignment allowed students to have meaningful conversations with their parents about soil and reinforced the relevance of science in their lives. (*The homework assignment highlights home culture connection to science by capitalizing on funds of knowledge from the students' homes and communities.*)

Ms. H. personally invited parents to stop in at any time and share their experiences about soil. She asked the school translator to make some of the calls. Mrs. Xiong, Chou's grandmother, came to the class to speak. Because she spoke Hmong, Mrs. Xiong spoke to the class through the school's interpreter. She compared the rich soil in Laos to the sandy soil in the Midwest and marveled that corn could grow at all in local soil, saying, "It's raining in Laos pretty much all the time so the soil is pretty much rich. It rains so much the forest holds everything together and holds the nutrients. It doesn't wash out. Over there we don't have sandy soil. In Wisconsin, I was so surprised to see corn growing in rows in the sandy soil. You walk in loose sandy soil. People say that sandy soil is not good soil, because plants like clay soil. Here, every year I thought the corn would never set, but it was good every season." The class had a lot of questions for Mrs. Xiong. They were interested in her comparisons of the fertilization techniques her family used in both countries. Some of the students were surprised to learn that there are worms in such a faraway country.

After sharing the parent interviews and hearing Mrs. Xiong's presentation, the class was convinced that soil was different in different places, but they wanted to be sure that this was true for soil from different places in their neighborhood, too. Ms. H. tried to center her science investigations in culturally relevant contexts, in this case their neighborhood. (*This "place-based" strategy established connections between school science and the students' community and lives.*)

Ms. H. encouraged students to gather physical evidence for their claim that "soil was different in different places." They decided that the best way to support their claim was to observe soil taken from different places near the school (practice: Planning and Carrying Out Investigations). They used a topographical map and an aerial photo map of the neighborhood to determine soil sites that seemed different: a hill, the marsh, and the school yard. They noticed that the sites had different trees—deciduous trees, no trees, and coniferous trees—and they also had different elevations (DCI: K-2-ESS2.B Earth's Systems). It was at these sites that the students collected and investigated the soil, digging pits and forming the basis for comparisons based on evidence and the soil profile diagrams each group constructed.

The following week, Ms. H. helped her students think in terms of patterns when exploring similarities and differences in the soil in the neighborhood (CC: Patterns). The students observed the soil color, texture, smell, and infiltration and collected data about the organisms in the soil. They learned a lot about patterns in soil composition (DCI: PS1.A Structure and Properties of Matter). Among other things, they learned that soil can be made up of sand, silt, clay, and organic materials, and that plants and animals are found in soil made with organic material. They discovered that sand has larger particles than silt and clay, and settles more quickly when it is in slowly moving water (CC: Energy and Matter). Some students were beginning to apply these ideas to the urban marsh soil with more organic material and coarser (sandy) mineral composition. The soil at the urban marsh had more sand than the other two sites due to rainwater flowing in from streets that were sanded during icy winter conditions. Trash, too, had collected in the urban marsh for similar reasons. The students witnessed trash blowing and collecting in the marsh during their fieldwork. Also, on the topographical map, they saw that the urban marsh was one of the lowest places in their neighborhood. Ms. H. wrote each piece of evidence on the whiteboard.

Ms. H. read the scientific question slowly and posted it on the board: "How do wind and water CHANGE the urban marsh soil?" (DCI: K-2-ESS2.A Earth's Systems). She used gestures while she spoke to show wind and water. After a "turn and talk" and a brief discussion, she used "stop the music" to get her students to talk to more than one partner about their thinking. The students moved around the room, and when the music stopped, they gathered in twos and threes and each shared. They had time to jot down whom they talked to and one thing they had discussed.

Jesus searched for words to explain how wind and water change the urban marsh soil, saying, "Water goes down, and because the coniferous had a hill and the urban marsh is down the hill and then the water. When people throw garbage in there and when the rain comes, it takes the garbage and put it in the urban marsh. And when the wind takes trash, when the wind comes, it throws the trash and it goes under and goes into the urban marsh." Pao responded, "Yeah, 'cause the wind blows the garbage." She showed what the wind did with her hands sweeping imaginary garbage.

Trinique was speaking with Moustafa and Kaleem. She said, "The sand is heavy and goes down in the water, not the clay! The clay floats away." Kaleem expanded on Trinique's idea, saying, "Because like the water can float, some stuff, that they can pick up stuff that are light and heavy and going down." Moustafa confidently added a new idea, "Yes, the wind move it too, garbage. The wind float it away" (CC: Stability and Change). The students used the concept of change to help understand and talk about the effect of the wind and water on the marsh soil.

All the students had time to write down and share their reasoning about the scientific question. Now they were ready to have a class discussion and felt comfortable expressing their ideas to the class.

The next day, Ms. H. asked the students, "How can we stop wind and rain from changing the urban marsh soil?" She asked the students to work on this question independently, first by drawing and then by explaining their drawing in writing (DCI: K-2-ETS1.B Engineering Design). When they worked independently, students came up with original

CHAPTER 9

ideas (Figures 9.4 and 9.5). Ms. H. noticed that many students tried to solve either the wind or the water problem, but not both. Pao drew a line of people holding up umbrellas over the urban marsh. Zytasia drew holes around the urban marsh that collected the garbage and sand. Moustafa thought that he could build a giant wall around the urban marsh. Victor drew a tent that completely covered the marsh. Kaleem considered an unusual solution; he sketched signs on the road, "Throw away your garbage!" and "Don't put sand on the road!" (practice: Constructing Explanations and Designing Solutions).

When the students finished their drawings and explanations of their designs, Ms. H. suggested "a museum walk" to share student ideas. She asked students to move around the room and use sticky notes to comment on each other's design presentations. Later, students went back to their own designs, read the comments, and refined their designs accordingly (DCI: K-2-ETS1.C Engineering Design).

FIGURE 9.4.

ENGINEERING SOLUTION 1

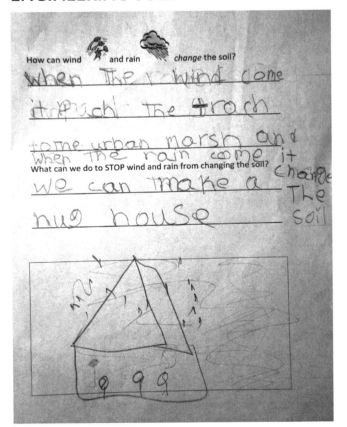

FIGURE 9.5.

ENGINEERING SOLUTION 2

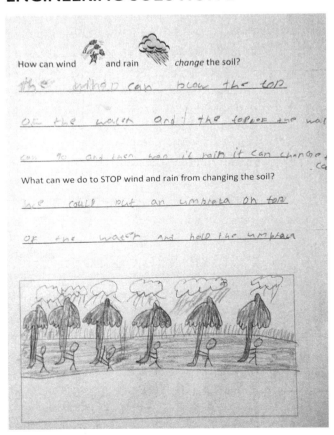

English Language Learners and the *Next Generation Science Standards*

NGSS CONNECTIONS

The *NGSS* promotes a vision of science learning by blending disciplinary core ideas, science and engineering practices, and crosscutting concepts. This vignette provides a snapshot of science practices and how disciplinary core ideas can be made accessible to English language learners by expanding on a crosscutting concept, Patterns, that the students may have encountered in other studies. See Figure 9.6 (p. 114) for the comprehensive list of *NGSS* and *CCSS* from the vignette. During the course of the six-week quarter (including lessons not described in this vignette), the students had opportunities to develop proficiency in all of the performance expectations in 2.ESS2, Earth's Surface System. The following performance expectations were highlighted in the vignette:

Performance Expectations

2-ESS2-1 Earth's Systems

Compare multiple solutions designed to slow or prevent wind or water from changing the shape of the land.

2-ESS2-2 Earth's Systems

Develop a model to represent the shapes and kinds of land and bodies of water in an area.

2-PS1-1 Matter and Its Interactions

Plan and conduct an investigation to describe and classify different kinds of materials by their observable properties.

K-2-ETS1 Engineering Design

Ask questions, make observations, and gather information about a situation people want to change to define a simple problem that can be solved through the development of a new or improved object or tool.

CHAPTER 9

Disciplinary Core Ideas

ESS2.A Earth Materials and Systems
Wind and water can change the shape of the land.
ESS2.B Plate Tectonics and Large-Scale System Interactions
Maps show where things are located. One can map the shapes and kinds of land and water in any area.
PS1.A Structure and Properties of Matter
Matter can be described and classified by its observable properties.
ETS1.C Optimizing the Design Solution
Because there is always more than one solution to a problem, it is useful to compare designs, test them, and discuss their strengths and weaknesses.

In the unit, the students developed understandings of core ideas in Earth science and physical science. Learning about the deposition of road sand in urban marsh in the familiar context became a catalyst for discussing water as a factor in the transportation of Earth materials in other settings. The students used their soil profile models to describe the layers of soil types, and classified the soil by observable properties. Also, students were challenged to design engineering solutions to mitigate road sand deposition in the marsh.

English Language Learners and the *Next Generation Science Standards*

Science and Engineering Practices

Developing and Using Models

Develop a model to represent patterns in the natural world.

Constructing Explanations and Designing Solutions

Make observations (first hand or from media) to construct an evidence-based account for natural phenomena.

Compare multiple solutions to a problem.

Planning and Carrying Out Investigations

Plan and conduct an investigation collaboratively to produce data to serve as a basis for evidence to answer a question.

Asking Questions and Defining Problems

Ask questions based on observations to find more information about the natural and/or designed worlds.

Define a simple problem that can be solved through the development of a new or improved object or tool.

During the six-week investigation, the students engaged in all eight of the science practices. This vignette highlighted the use of Developing and Using Models. The students refined and expanded their soil profile models to explore patterns as they developed an understanding of Earth's surface systems. The teacher provided various entry points for all students to engage in the science practice of Constructing Explanations and Designing Solutions. As students constructed their explanations, she relied on collaborative group activities to help students gather evidence and make claims based on the evidence. She also included individual writing and presentations to design engineering solutions, and the students applied this practice in multiple formats. Finally, the students *planned and carried out an investigation* with support, centered on gathering evidence to address the question, "Is all soil the same?"

CHAPTER 9

Crosscutting Concepts

Patterns

Patterns in a natural world can be observed.

Stability and Change

Some things stay the same while other things change.

Energy and Matter

Objects may break into smaller pieces and be put together into larger pieces, change shapes.

Structure and Function

The shape and stability of structures of natural and designed objects are realted to their function(s).

The crosscutting concepts of Patterns and Stability and Change are overarching ideas in this unit, and Energy and Matter was a focus during the investigations of the physical properties of soil. The teacher encouraged the students to think about their scientific experiences with soil in terms of patterns when she posed questions about the similarities and differences in the soil in their neighborhood. The students noticed a pattern in the prevalence of organisms in certain types of soil, and noted the presence of worms, insects, and other living creatures in some soil types but not in others. They observed that water filtration was also connected to the soil type. They also found a pattern in the amount of organic material in the soil and its elevation. The students learned that changes, as in the slow changes indicated by the soil layers and faster changes caused by sand deposits on the road, are a valuable way to think about soil. In this way, Ms. H.'s students gradually developed an understanding of how the crosscutting concepts of patterns, and stability and change, apply to Earth science.

CCSS CONNECTIONS TO ENGLISH LANGUAGE ARTS AND MATHEMATICS

The *NGSS* is committed to addressing the *Common Core State Standards* in English language arts (ELA) and mathematics as an integrated part of science. English language learners

English Language Learners and the *Next Generation Science Standards*

develop mastery of the language of ELA and math when they engage in these subject areas in a meaningful context and for an authentic purpose. The vignette highlighted the teacher's incorporation of the *CCSS ELA*. Within the larger context of Earth science, the students collaborated to problem solve and develop claims, evidence, and reasoning in small and large groups.

- **SL.2.1** *Participate in collaborative conversations with diverse partners about grade 2 topics and texts with peers and adults in small and large groups.*

Students also addressed the CCSS for math as part of their evidence gathering. The unit repeatedly focused on this standard in preparation for and part of data collection and development of the soil profile models.

- **2MD.1** *Measure and estimate lengths in standard units: Measure the length of an object by selecting and using appropriate tools such as rulers, yardsticks, meter sticks, and measuring tapes.*

EFFECTIVE STRATEGIES FROM RESEARCH LITERATURE

The *NGSS* science and engineering practices are language intensive and require students to engage in classroom science discourse. Students must read, write, view, and visually represent as they develop their models and explanations of scientific phenomena. They speak and listen as they present their ideas or engage in reasoned argumentation with others to refine their ideas and reach shared conclusions. These practices offer rich opportunities and demands for language learning at the same time as they support science learning (Lee, Quinn, and Valdés 2013). The literature indicates five areas where teachers can support science and language for English language learners: (1) literacy strategies with all students, (2) language support strategies with English language learners, (3) discourse strategies with English language learners, (4) home language support, and (5) home culture connections (Fathman and Crowther 2006; Lee and Buxton 2013; Rosebery and Warren 2008).

First, teachers highlight various strategies for literacy development (reading and writing), such as activating prior knowledge, having explicit discussion of reading strategies for scientific texts, prompting students to use academic language functions (e.g., *describe, explain, predict, infer, conclude*) for science and engineering, engaging students in scientific genres of writing (e.g., keeping a science journal), teaching the uses of graphic organizers (e.g., concept map, word wall, Venn diagram), and encouraging reading trade books or literature with scientific themes.

Second, teachers provide language support strategies with English language learners, typically identified as English for Speakers of Other Languages (ESOL) strategies. They use hands-on activities, realia (real objects or events), and multiple modes of representation

CHAPTER 9

FIGURE 9.6.
NGSS AND CCSS FROM VIGNETTE

K-2-ETS1 Engineering Design

2. Earth's Surface Systems: Processes that shape the Earth

2-PS1 Matter and its Interactions

Students who demonstrate understanding can:

2-ESS2-1. Compare multiple solutions designed to slow or prevent wind and water from changing the shape of the land.

2-ESS2-2. Develop a model to represent the shapes and kinds of land and bodies of water in an area.

2-PS1-1. Plan and conduct an investigation to describe and classify different kinds of materials by their observable properties.

K-2-ETS1-1. Ask questions, make observations, and gather information about a situation people want to change to define a simple problem that can be solved through the development of a new or improved object or tool.

The performance expectations above were developed using the following elements from the NRC document *A Framework for K–12 Science Education*:

SCIENCE AND ENGINEERING PRACTICES	DISCIPLINARY CORE IDEAS	CROSSCUTTING CONCEPTS
Developing and Using Models Modeling in K–2 builds on prior experiences and progresses to include using, and developing models that represent concrete objects or design solutions. • Develop a model to represent patterns in the natural world. **Constructing Explanations and Designing Solutions** Constructing explanations and designing solutions in K–2 builds on prior experiences and progresses to the use of evidence or ideas in constructing explanations and designing solutions. • Compare multiple solutions to a problem. **Planning and Carrying Out Investigations** Planning and carrying out investigations to answer questions or test solutions to problems in K–2 builds on prior experiences and progresses to simple investigations, based on fair tests, which provide data to support explanations or design solutions. • Plan and conduct an investigation collaboratively to produce data to serve as the basis for evidence to answer a question.	**ESS2.A: Earth Materials and Systems** • Wind and water can change the shape of the land. **ESS2.B: Plate Tectonics and Large-Scale System Interactions** • Maps show where things are located. One can map the shapes and kinds of land and water in any area. **ETS1.C: Optimizing the Design Solution** • Because there is always more than one solution to a problem, it is useful to compare designs, test them, and discuss their strengths and weaknesses. **PS1.A: Structure and Properties of Matter** • Different kinds of matter exist and many of them can be either solid or liquid, depending on temperature. Matter can be described and classified by its observable properties.	**Patterns** • Patterns in the natural world can be observed. **Stability and Change** • Some things stay the same while other things change. **Energy and Matter** • Objects may break into smaller pieces and be put together into larger pieces, or change shapes. **Structure and Function** • The shape and stability of structures of natural and designed objects are related to their function(s).

CCSS Connections for ELA and Mathematics

SL.2.1 Participate in collaborative conversations with diverse partners about grade 2 topics and texts with peers and adults in small and large groups.

2MD.1 Measure and estimate lengths in standard units: Measure the length of an object by selecting and using appropriate tools such as rulers, yardsticks, meter sticks, and measuring tapes.

English Language Learners and the *Next Generation Science Standards*

(gestural, oral, pictorial, graphic, textual). They guide students to comprehend key science vocabulary in context—both general academic terms and discipline-specific terms.

Third, discourse strategies focus specifically on the teacher's role in facilitating English language learners' participation in classroom discussion to enhance their understanding of academic content (i.e., adjust the level and mode of communication). A major challenge for teachers is in how to structure activities so as to reduce the language barrier for participation while maintaining the rigor of science content and processes.

Fourth, teachers can build upon and make use of students' home language to support science learning in English. Teachers may introduce key science terminology in both the home language and English, highlight cognates as well as false cognates between English and the home language, allow code-switching, and encourage bilingual students to assist less English proficient students in home language.

Finally, to connect science to students' home culture, teachers need to understand that students participate in classroom interactions in ways that reflect culturally based communication and interaction patterns from their home and community. In addition, they need to elicit students' "funds of knowledge" related to science topics and use students' cultural artifacts and community resources in ways that are academically meaningful and culturally relevant.

CONTEXT

DEMOGRAPHICS

The number of school-age children (ages 5–17) who spoke a language other than English at home rose from 4.7 million to 11.2 million between 1980 and 2009, or from 10% to 21% of the population in this age range (NCES 2011). Currently, more than one in five students (21%) speak a language other than English at home, and limited English Proficient (LEP) students (the federal term) have more than doubled from 5% in 1993 to 11% in 2007. This statistic does not include students who were classified as English language learners when younger but who are now considered fluent English speakers.

Although Spanish speakers make up the majority of the ELL population in the United States, there are over 400 different languages spoken by U.S. students (U.S. Census Bureau 2012). In 2009, of the total number of students ages 5–17 who spoke a language other than English in the home or spoke English with difficulty, 73.5% spoke Spanish, 10.5% spoke an Indo-European language, 12.6% spoke an Asian or Pacific Islander language, and 3.5% spoke other languages (U.S. Census Bureau 2011). Public schools in the nation's urban areas have more English language learners (14% on average) than schools in suburban areas (8%), towns (7%), or rural areas (4%) (NCES 2009). English language learners

CHAPTER 9

comprise foreign-born and U.S. citizens; 80% of elementary school English language learners are born in the United States (IU Newsroom 2008).

SCIENCE ACHIEVEMENT

Science achievement of English language learners continues to fall behind non-English language learners. For example, based on the 2009 National Assessment of Educational Progress (NAEP) science scores, only 3% of ELL eighth graders scored proficient on the science achievement assessment, compared to 34% of non-English language learners. In the same year, 17% of eighth-grade English language learners scored basic or above, compared to 68% of non–English language learners (NAEP 2011). The science achievement gaps between ELL and non–English language learners widened considerably from 2005 to 2009 for 4th, 8th, and 12th graders. According to the 2009 NAEP science results using 300-point scale scores, the gaps were 40 points for 4th graders, 50 points for 8th graders, and 47 points for 12th graders. These widening gaps were a reversal of the trend from the previous decade, in which gaps narrowed somewhat from 1996 to 2005 for 4th and 8th graders. NAEP does not break down ELL science scores by language group, type of district (urban, suburban, towns, or rural), or origin (U.S. vs. foreign-born).

Before 1996, NAEP did not offer any accommodations for English language learners and students who could not meaningfully participate in the assessment were excluded (NCES 2013). The NAEP science assessments conducted in 1996, 2000, 2005, and 2009 included accommodations such as extended time, test items read aloud in English or Spanish, translated assessments, and bilingual dictionaries and glossaries. Unfortunately, the lack of consistent policy regarding the participation of English language learners produced wide variability in exclusion rates making it difficult to meaningfully assess ELL growth. In 2010, in an effort to be more inclusive, NAEP implemented a new policy that extended accommodations to all English language learners and enforced higher (85%) inclusion rates.

EDUCATION POLICY

Part A of Title III of the Elementary and Secondary Education Act, the English Language Acquisition, Language Enhancement, and Academic Achievement Act, provides for monitoring English language learners for Adequate Yearly Progress (AYP) in English language proficiency and content areas. This act emphasizes increased accountability of English language learners in content-based academics and delivers funding according to each school's ability to meet AYP.

To measure progress, each state is required to implement English language proficiency tests based on its English language proficiency standards, which in turn are linked to the state's academic standards. Proficiency levels (usually ranging from *limited proficiency* to *proficient*) must be established to measure progress; however, these levels are not consistently defined across states. Even though 28 states have adopted World Class Instructional Design from the WIDA Consortium to monitor progress, there is little consistency nationwide. As definitions of ELL, goals for ELL achievement (AYP), and monitoring tools vary, it is impossible to compare ELL performance across states.

Content area assessment and measures of AYP for English language learners cannot be separated from the language programs that serve students, also dictated by state policy. If a state supports bilingual education, then it is likely that at least some portion of instruction is conducted in the students' home language while they are developing academic language proficiency in English. In states that follow an "English only" policy for English language learners, then all science instruction takes place in their second language and science knowledge has to be developed concurrently with academic English. Overall, educational policy for English language learners has been moving away from bilingual education and multicultural perspectives, and toward English proficiency and accountability for standards-based content learning.

REFERENCES

Fathman, A. K., and D. T. Crowther, eds. 2006. *Science for English language learners: K–12 classroom strategies.* Arlington, VA: National Science Teachers Association.

IU Newsroom. 2008. Advocate for English language learners and immigrant rights headlines school of education program. Indiana University news room (June 25). *http://newsinfo.iu.edu/news/page/normal/8470.html*

Lee, O., and C. A. Buxton. 2013. Integrating science learning and English language development for English language learners. *Theory Into Practice* 52 (1): 36–42.

Lee, O., H. Quinn, and G. Valdés. 2013. Science and language for English language learners: Language demands and opportunities in relation to *Next Generation Science Standards*. *Educational Researcher* 42 (4): 223–233.

National Center for Education Statistics (NCES). 2011. *The condition of education 2011* (NCES 2011-033). Washington, DC: U.S. Department of Education.

National Center for Education Statistics (NCES). 2013. National Assessment of Educational Progress. *http://nces.ed.gov/nationsreportcard/about/inclusion.asp*

Rosebery, A. S., and B. Warren, eds. 2008. *Teaching science to English language learners: Building on students' strengths.* Arlington, VA: National Science Teachers Association.

U.S. Census Bureau. 2012. *Statistical abstract of the United States, 2012.* Washington, DC: U.S. Government Printing Office. *www.census.gov/compendia/statab/cats/education.html*

CHAPTER 10

GIRLS AND THE *NEXT GENERATION SCIENCE STANDARDS*

MEMBERS OF THE *NGSS* DIVERSITY AND EQUITY TEAM

ABSTRACT

Despite gains in science achievement scores, girls lag behind boys in every grade tested in the National Assessment of Educational Progress (NAEP). By the time girls reach high school, a disproportionate number steer away from advanced courses in science—physics, engineering and computer technology—limiting their options for STEM (science, technology, engineering, and mathematics) college degrees or careers. Research points to three main areas where schools can positively impact girls' achievement, confidence, and affinity with science: (1) instructional strategies, (2) curricular decisions, and (3) classroom and school structure. The *Next Generation Science Standards* (*NGSS*) pave the way for increased exposure to all disciplines of science for all students. This is a breakthrough, in particular, for girls, as research attributes gender disparities in science achievement, college graduation, and career success to an early "experience gap" between girls and boys. The vignette below highlights an early exposure to engineering through a forest restoration project that girls found engaging. It underscores how the purposeful inclusion of effective strategies for girls can have a positive impact on their confidence as beginning scientists and engineers.

VIGNETTE: DEFINING PROBLEMS WITH MULTIPLE SOLUTIONS WITHIN AN ECOSYSTEM

While the vignette presents real classroom experiences of the *NGSS* implementation with diverse student groups, some considerations should be kept in mind. First, for the purpose of illustration only, the vignette is focused on a limited number of performance expectations. It should not be viewed as showing all instruction necessary to prepare students to fully understand these performance expectations. Neither does it indicate that the performance expectations should be taught one at a time. Second, science instruction should take into account that student understanding builds over time and that some topics or ideas require extended revisiting through the course of a year. Performance expectations will be realized by using coherent connections among disciplinary core ideas, science and

CHAPTER 10

engineering practices, and crosscutting concepts within the *NGSS*. Finally, the vignette is intended to illustrate specific contexts. It is not meant to imply that students fit solely into one demographic subgroup, but rather it is intended to illustrate practical strategies to engage all students in the *NGSS*.

INTRODUCTION

The following vignette highlights an example from each of the three main areas—teaching strategies, curricula, and organizational structure—that raise achievement and confidence in science for girls. The unit described in this vignette (based on Earth Partnership for Schools Program 2010) illustrates how outcomes for engineering practices can be achieved through a project that young girls find engaging. Engineering, practiced by fewer females than males, has a particular appeal for the girls in this vignette because the practices are developed around the core content in life sciences, a discipline with which more girls tend to identify.

Although the unit—improving habitats for forest wildlife—focuses on the natural world rather than the designed world, the practice falls squarely within the realm of engineering (solving a problem) more than science (answering a question). The vignette begins after a unit on ecosystems to meet the performance expectations for "Matter and Energy in Organisms and Ecosystems." Recognizing that the students needed to be more engaged in thinking about the importance of ecosystems, Ms. G., the teacher, extended the ecosystem unit using additional activities to meet specific performance expectations related to "Engineering Design."

Through the course of the unit, many of the girls in the class are transformed from passive observers to active participants solving a real world problem. As a result of careful planning on the part of their teacher, the girls proudly call themselves "engineers" and choose to take on leadership roles. Ms. G. uses the project-based inquiry model as well as literature and design to spark her girls' interest in engineering. Ms. G. additionally encourages the girls to join the after-school science club, allows them to form all-girl teams in science, and reinforces the message that science is for everyone. In the vignette, Cassidy, Aurelia, Annika, and Kailey use science practices and content to identify, assess, and solve a problem that has important real-world consequences. Through this experience, they gain an understanding of the application of science toward engineering solutions, and thereby establish a foothold in the world of engineering. Throughout the vignette, classroom strategies that are effective for all students—particularly for girls according to the research literature—are highlighted in parentheses.

CONNECTIONS TO GIRLS IN SCIENCE AND ENGINEERING

Engineering: The Solution

Kailey, a third grader, stood ready to speak in front of the entire third grade of her school. Three classrooms of students were sitting cross-legged on the floor of the school's cafeteria. She and her classmates were about to present their forest management problem and its solution. Kailey was the student in her class most eager to present. Annika, Aurelia, Billy, and Josh were next to her, more shy but equally excited about the project.

FIGURE 10.1.

PHENOLOGY WHEEL PRESENTATION

After a few words from the teacher, Kailey stepped forward. She began by stating the question: "Do our woods give the animals enough food?" Next to her was a large colorful phenology wheel (Forbes 2008) that showed the evidence that the class had assembled (Figure 10.1).

Kailey read from the cue cards she and her class had prepared: "In the fall, our class did a habitat activity in the woods." An appreciative murmur sounded in the cafeteria; many students had enjoyed activities in the woods. "We counted trees, shrubs, flowers, and old logs. We counted dens and nests. We were looking for signs of food and shelter, things that all forest animals need in their home, or habitat. We wanted to see where food comes from in our woods."

Kailey passed the cue cards to Josh. He continued, listing and describing the food sources in the school woods, including seeds, fungi, insects, leaves, and smaller creatures eaten as prey. Next, Annika spoke: "After we learned all the kinds of food in our woods, we wanted to know in what months that kind of food was available. So we made this calendar-wheel, which shows the months of the year. You can see on this wheel that some foods, like fungi and meat from creatures, are available in the woods all year."

Annika took a breath and remembered to slow down. "You can also see on our wheel that there are white spaces. The white spaces show when food is not available; so this ring shows that nectar from flowers is available only in April, May, and June, but not in July, August, September, or October. Since every white space shows a time when food is not available, some animals might not have enough food in those months. We decided to try to make our woods a better home for animals."

Aurelia held up a large picture of a white flower, and Annika handed Billy the cue cards. He stepped forward to finish the presentation, "We learned about a native woodland flower, the arrow-leaf aster that blooms in the late summer and all fall until the frost comes. This will make the flower nectar available for bees and other pollinators for a much longer time. So we thought it would be good to plant it in our woods. And the good news is we can! Ms. G. ordered enough plants from a native plant nursery so each class can plant some." Billy glanced up at the crowd and added, "Next year, in the fall, we will check to see if our plan worked by doing a flower observation and looking to see if bees are using the flowers."

The third grade applauded enthusiastically at the close of the presentation, and Annika, Aurelia, Josh, Billy, and Kailey stood proudly next to their classmates. The effort to enhance the school woods was on its way to becoming a success. Their teacher, Ms. G., was very proud of her students and reflected on how far they had progressed through the project activities.

Planning an Investigation

Ms. G. was a seasoned teacher with a passion for getting students outside. She took her students to the nearby school woods weekly or biweekly and tried to figure out creative ways to meet the science standards through the lens of the woods. She found that the woods and the outdoors engaged her students because they could see the science. This year, she was determined to achieve the engineering objectives in the school woods as well. Through this effort, she knew the woods study would engage the girls as much as the boys, and she would enlist her students as lead partners in tackling an engineering design problem.

Ms. G. had a group of 20 third graders, 12 girls and 8 boys. About half of the students lived in the town of 1,400 and in outlying subdivisions and the rest lived on farms and arrived on buses every morning. Over the years, she found that there were a few students in each class who came to school having explored natural areas on their own. When the class went outside, these students tended to notice more than the other students about their natural surroundings. However, as the year progressed, all of her students could increasingly describe the woods, its seasonal changes, and the interdependence of living things in the wild. Through engineering design, she planned to engage her students as active participants in the ecosystem investigation.

She began the engineering project by setting the stage for her students to identify an engineering problem. She proposed a question for investigation: "Are our woods a good home for wildlife?" (*Ms. G. chose a curriculum topic that had relevancy and real-world application, an angle that would purposefully interest and engage the girls in the class.*)

Her students, sitting in desks facing the whiteboard, were silent. Ms. G. waited. After a minute, Aurelia raised her hand. Her family had a CSA (community supported agriculture) farm, and she was very knowledgeable about plants. Billy, another confident student, raised his hand next. He spent hours out in the country hunting with his father. Ms. G. waited another minute as a few more hands went up. She acknowledged the students who wanted to share and nodded to Aurelia.

"Ms. G., I think our woods are a good home for the animals! The deer … raccoons, chipmunks, and lots of birds … like cardinals and woodpeckers … there is pretty much nature.

Girls and the *Next Generation Science Standards*

Last Tuesday when we were pulling out the buckthorn, we saw that fat toad, and remember?" She looked around and dreamily said, "He was so *happy*." Some of the students laughed; that toad *had* seemed extremely pleased with himself. Ms. G. laughed too, and then called on Billy. "I think so, too. The woods are a great home," he continued. "Animals aren't like people. They are okay with a hard life, even, because that's what they're used to. They just run and survive in the woods. If they are *there*, then it's a good home, if they have enough food." Abby spoke up with a question, "How do we know for sure? I'm wondering: If it's such a great home, why don't we see a lot more animals?" (practice: Asking Questions and Defining Problems). Billy said, "We know that they are hiding, and that's why we don't see them! Animals are really good at hiding and blending in."

After a few more responses, and a few animal anecdotes, Ms. G. said, "I want to go back to Abby's question, "How *do* we know for sure that our woods are a good home for the animals? How could we find out? Today, when we walk in the woods, I want you to think like wildlife biologists. Does anyone know what a wildlife biologist does?" Ms. G. waited, but this time no one ventured a guess. She asked again, "Does anyone know what a wildlife biologist might *look for* in our woods?" Again, the class was silent, and Ms. G. tried another approach, "Remember our visitor, Ms. DeNotter? She is studying to be a herpetologist, a reptile expert, which is a kind of wildlife biologist." (*Ms. G. previously had brought in a female expert in science, a strategy that enhances girls' confidence in science.*) "What do you think wildlife biologists care about?"

A chorus of students responded together, "Animals!" and "Wildlife!" Ms. G. continued, "So, in our woods, what would a wildlife biologist care about?" Kailey—who had written an effusive thank-you letter to Ms. DeNotter—raised her hand to say, "Well, animals need food, and shelter, and they need safe places to hide." Ms. G. responded, "Kailey, on our walk today, as wildlife biologists, what might we look for?" Kailey thought for a second, and then said, "I think we will be looking for animal homes and what foods the animals eat out there." She looked over at Aurelia who nodded (practice: Planning and Carrying Out Investigations). The class generated a list of animal foods and homes in the woods to write down in their science notebooks. The "food" list included nuts, berries, fungi, leaves, bugs for birds, and little animals; the "homes" list had nests, burrows, dens, holes in logs, and rotten logs.

Ms. G. took out her oak tree picture and put it on the overhead. "Why might it be important to a wildlife biologist to know how many oak trees there are in our woods? Talk to the person next to you." (*Ms. G. carefully planned the partners and groupings in her class, an organizational structure that encourages participation for the girls in science.*) After the pairs of students talked, Ms. G. called on Grace and Megan. Megan deferred to Grace, who had a little more outdoor experience but rarely spoke. Grace responded, "Well, more oak trees would make more acorns for squirrels to eat."

"Nicely said!" Ms. G. was pleased. "Why do you think it might be important for a wildlife biologist to look up into the canopy and the understory as well as on the forest floor?

CHAPTER 10

FIGURE 10.2.
MAP OF WOODS

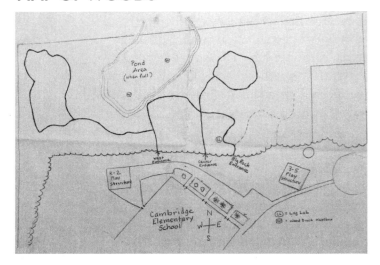

FIGURE 10.3.
FOOD SOURCE INVENTORY

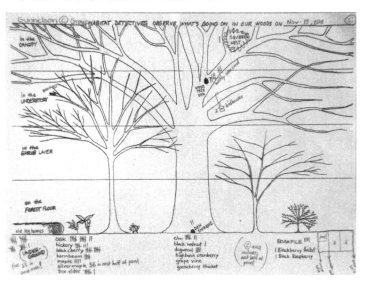

Annika?" Annika had a big garden at home and also spent time hiking in the forest with her mom. She answered, "Because berries might be up there, or a nest or something, but down here you could find acorns or a hole in a log." Ms. G. responded, "Got it, thanks! Who else wants to add to Annika's response?"

Ms. G. put the hand-drawn map of the woods (Figure 10.2) on the overhead projector and told the class, "Today we're going off the trail. We are going to be responsible for this section of our woods, looking at it as a wildlife biologist would … so let's all remember: We are looking for clues that tell us this is a good habitat, a good home for our wildlife."

Analyzing and Interpreting Data

After the students had collected and tallied observations of the woods, Ms. G. needed her students to become familiar with their data. Ms. G. placed her own large map on the whiteboard next to the three vocabulary words *Quantity*, *Space*, and *Time*, and she showed the students how to explore the data (Figure 10.3), saying, "We can look at the data we collected in different ways, 'quantitatively' or *how many* of a plant species there are, and 'spatially,' meaning *where* they are. Think about our walk in the woods and think about what things you noticed quantitatively and spatially about our woods." She checked for responses by saying, "I would like you to work with your partner and notice something spatial and something quantitative, and write it down in your science notebook. After you get one or two ideas on your paper, you can come up and write an 'observation' on the chart paper" (practice: Analyzing and Interpreting Data).

Girls and the *Next Generation Science Standards*

In a few minutes, Josh and Billy came to the front of the room and recorded their question on the chart paper: "How come all the silver maples live around the pond?" Aurelia and Kailey wrote, "There are more box elders in the woods than any other trees." "There are HARDLY ANY black walnut trees—only 3 in the whole woods," came from another group. And Abby's group wrote, "Most of the dogwood is in the center part of our woods." Grace and Megan added, "Mushrooms are everywhere!"

Using Modeling to Define and Refine an Engineering Problem

Next, Ms. G. took out the phenology wheel she had prepared for the lesson. Annika saw the wheel and whispered to a few others, "Phenology wheel!" She had loved the two phenology wheel activities Ms. G. had introduced previously: a class birthday activity and "enhancing a sense of place" sit spots. Ms. G.'s next goal was to make a model to explain the food availability using the phenology wheel.

Ms. G. stated, "We looked at our data spatially and quantitatively. Now we need to add the dimension of *time*. This phenology wheel is designed as a circular *calendar*, with a ring for each food type." She passed out a science folder for each pair of students that contained phenology observations from the woods from different years. She said, "Using the past two years of phenology observations that we have, along with your notes, we can determine the dates that these foods are available in our woods. We plan a color scheme that will tell the story of our data (Figure 10.4), and then we color the wheel that color on the months the food is in the woods (Figure 10.5). So if we do not color part of a ring, then it means *none of the* food can be found during that month (Figure 10.6, p. 126). What kind of food do you think should go with the color green?"

FIGURE 10.4.

INVENTORY AND COLOR

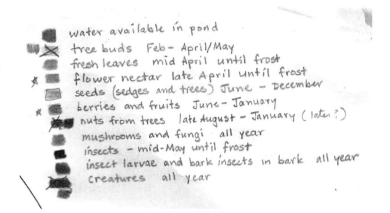

FIGURE 10.5.

STUDENTS WORKING ON PHENOLOGY WHEEL

CHAPTER 10

FIGURE 10.6.

PHENOLOGY WHEEL WITH WHITE SPACES

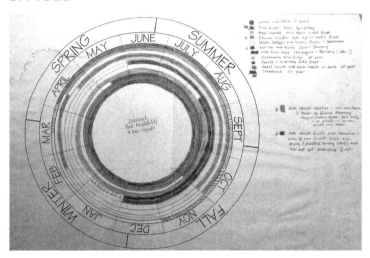

Figuring out the data and accurate coloring of the phenology wheel was a challenging undertaking and called for a lot of group collaboration. The partners circulated and shared the notes about the first and last dates a flower or leafy plant or fungi was observed in the woods. When it was finally finished, each student looked at the colorful wheel. The class was surprised at the results. Josh exclaimed, "Hey, you guys, look at all the white spaces!" Billy nodded knowingly: "You know what the white means; it's hard times." "Hard times! I know it," Josh answered. Kailey was distressed as she noted, "White is when they are hungry …" and then added, "Can we feed them?" (practice: Asking Questions and Defining Problems).

A discussion ensued about how to phrase the problem. The students thought that the problem should be stated as follows: "Problem: In some months, there might not be enough food (berries, nectar, leaves) for all the wildlife."

Multiple Design Solutions to an Engineering Problem

The class discussed two possible solutions right away. Annika thought they should put nuts and seeds out in the woods like she did in her yard in the winter for squirrels and cardinals. A few students agreed, but Billy reflected that a lot of animals buried the nuts. He thought the wheel should show the nuts were underground, and he didn't think nuts were the biggest worry. He pointed out animals eat different food in the winter, like deer eat bark and twigs when there are no more leaves. Josh reminded everyone that a lot of animals, such as birds, depend on berries for food. He was troubled about there being no berries until July in the woods (practice: Engaging in Argument From Evidence).

Grace offered to ask her parents to donate fruit and berries for the animals. Dubiously, Aurelia commented, "Ms. G., you have to feed them all the *time*. You can never forget or else … the animals depend on you to *bring* it. What will they do next year?" Aurelia looked at Grace and said, "When we are in fifth grade, even more animals will starve." Kailey carefully chose her words, responding, "The wheel shows food from the woods—food that always grows in the woods. Can't we plant something that will always feed the animals in these months?" She indicated the white spaces on the wheel and said, "That would help the animals" (DCI: ETS1.A and ETS1.B Engineering Design).

Girls and the *Next Generation Science Standards*

Ms. G. explained that changing a habitat to solve a problem is engineering design, called *habitat management* or *restoration*. She explained engineers use models, like the phenology wheel, to get a sharp look at their problem. They could use the phenology wheel to figure out what things in the habitat might need to change to provide for animals during the months when their food is not abundant. It is this feature, the phenology wheel's usefulness in defining the problem, which makes it a model. Ms. G. helped the class begin to think about habitat improvement by looking at each food ring. It was easy to see they couldn't do much about rain, snow, and the pond. Nor could they command the leaves to unfurl during the months when no leaves were out. Ms. G. instructed each pair to examine their phenology wheels and come up with some suggestions (CC: Cause and Effect). (*Having students generate the problems and possible solutions that will drive class activities, an important scientific practice, is also very motivating for all students, including girls.*)

Abby immediately addressed the class when they reconvened: "What about flowers? Look at how they bloom only from April through June. A lot of insects need to eat nectar. Could our class do something about the flowers?" Annika, Abby's partner, said, "There are lots of flowers that grow after June, like in July and even in the fall in my mom's garden. We have lilies, and chrysanthemums, and sunflowers. We could plant those flowers in the woods, but my mom's garden has a lot of sun. Can they grow in our woods?" (practice: Asking Questions and Defining Problems).

Ms. G. commented that the criterion would be important to consider if they made a decision to plant something in the woods. She wrote the word *criterion* and the definition on the board: "The information we need to have before choosing our solution." She made a grid on the whiteboard. Under the heading "Possible Habitat Solutions" Ms. G. wrote: "1. Plant—late flowering plant, food source: nectar." She made four columns under "criteria" and wrote, "Will it grow in our woodland?" in the first column. She left the next three columns blank. She explained they would be considering other important criteria for their engineering solution and writing them in the remaining columns with Ms. I, a volunteer community naturalist.

Although Ms. G. loved the woods, she was a novice. So she found an expert who could answer her questions. Ms. I. oversaw a weekly after-school club dedicated to doing activities to take care of the woods. (*The schools' after-school club represents a structural decision that positively impacts girls and their attitude toward science.*) Ms. G. encouraged all of her students to join. Recently the club had worked to remove garlic mustard, buckthorn, and honeysuckle. The club was well run and Ms. I. was loved by staff and students.

Grace said, without being called on, "What do the bees do when the woods are all out of flowers?" Ms. G. smiled encouragingly and said, "Good question! I actually don't know the answer." She asked Grace to write the question down on the "Science questions" board by the word wall. She continued, "Annika and Abby wondered if there are some native late-blooming woodland flowers and if planting those might be a great solution to our engineering problem. We will definitely have to do some research" (DCI: 3-5 ETS1 Engineering Design).

CHAPTER 10

After the class looked carefully at the phenology wheel, they expanded their list to address their concerns. The list under "Habitat Solutions" expanded to include:

1. Plant—late flowering plant, food source: nectar
2. Plant— early fruiting berry bushes, food source: berries
3. Plant—more black walnut trees, food source: nuts

Coming up with the most important and appropriate criteria was more difficult, and Ms. G. invited Ms. I. to assist the class on this step. Ms. I. helped the students define five considerations to keep in mind as criteria for making their decision. With her guidance, the class modified the first criterion slightly and then added four new criteria to the grid:

1. Will the plant grow in an oak-hickory forest?
2. Is it native?
3. Will it help wildlife?
4. Can we afford it?
5. Are there any negative effects from planting it?

Ms. I. also suggested that the students research some plants according to their newly established criteria. She proposed that they could research three plants: the arrow-leaf aster, the serviceberry, and the black walnut tree. She said that although each plant was a potential solution to the problem, they needed to carefully consider all of the criteria before deciding on a solution (CC: Cause and Effect).

After a week of researching native plant guides and websites and making a few calls to a plant nursery, the students evaluated each solution to the forest engineering problem. They developed a number system to rate how well each solution met the criteria. They could then evaluate the possible solutions, side-by-side, putting a number from 1 (does NOT meet this criterion) through 4 (meets this criterion well) in each space in the grid. The first criterion was, "Will the plant grow in an oak hickory forest?" (3-LS4 LS4.C: Adaptation).

All three plants grew in this type of forest (criterion 1) and they were all native (criterion 2), so the remaining three criteria became most critical for deciding which species to plant (practice: Constructing Explanations and Designing Solutions).

The serviceberry would feed the most species (insects, birds, chipmunks, and squirrels) during crucial times, because it produced flowers and berries in the slim months. Although the cost for this shrub was reasonable, it cost much more than the asters, so it was assigned a 2 for "Can we afford it?"

Girls and the *Next Generation Science Standards*

The black walnut was problematic. The students weren't convinced the animals did not already have access to nuts year-round. Also, Abby and Annika found evidence that walnut trees produce chemicals that make it hard for some plants to grow.

Planting asters caused the least harm to the woods' floor because only small holes needed to be dug for their roots. However, asters wouldn't be able to feed as many species. The serviceberry got the most points, and the decision to plant serviceberry was just about unanimous.

Kailey, however, was still unconvinced about serviceberry. She was sure that the asters were a better solution. She noted that some criteria mattered more to her than other criteria (CC: Influence of Science, Engineering and Technology on Society and the Natural World). Kailey said, "They aren't all the same, you know? Me and Aurelia think that they should not be the same number of points. If we can only plant one serviceberry with our money, it won't be enough to make a real difference for the animals!" Aurelia chimed in, "Also, you know what else? We didn't give the asters points for feeding birds. It should be three points, because birds eat the insects." Annika changed her vote to asters and asked for another class vote. Their argument was convincing. The majority of the class voted to plant asters (practice: Engaging in Argument From Evidence).

Ms. G.'s students were excited and confident in their decision. They had used data to generate evidence, models to help them predict, and engineering practices to solve a problem in the world around them. Through the process Ms. G's students had gained a deep understanding of their woods. They were now engineers and scientists ready to continue the engineering work of improving their woods.

NGSS CONNECTIONS

Each *NGSS* performance expectation applies equal weight to the three dimensions of disciplinary core ideas, science and engineering practices, and crosscutting concepts. See Figure 10.7 (p. 135) for the comprehensive list of *NGSS* and *CCSS* from the vignette.

CHAPTER 10

Performance Expectations

3-5-ETS1-1 Engineering Design

Define a simple design problem reflecting a need or a want that includes specified criteria for success and constraints on materials, time or cost.

3-5-ETS1-2 Engineering Design

Generate and compare multiple possible solutions to a problem based on how well each is likely to meet the criteria and constraints of the problem.

3-LS4 Biological Evolution: Unity and Diversity

Construct an argument with evidence that in a particular habitat some organisms can survive well, some survive less well, and some cannot survive at all.

The class activities described in this vignette helped students make progress toward the performance expectations. In the vignette, the students blended core ideas about engineering with several different science practices, and applied an understanding of the crosscutting concept of how science, engineering, and technology influence society and the natural world.

Disciplinary Core Ideas

ETS1.A Defining and Delimiting Engineering Problems

Possible solutions to a problem are limited by available materials and resources (constraints). The success of a designed solution is determined by considering the desired features of a solution (criteria). Different proposals for solutions can be compared on the basis of how well each one meets the specified criteria for success or how well each takes the constraints into account.

ETS1.B Developing Possible Solutions

Research on a problem should be carried out before beginning to design a solution. Testing a solution involves investigating how well it performs under a range of likely conditions.

At whatever stage, communicating with peers about proposed solutions is an important part of the design process, and shared ideas can lead to improved designs.

LS4.C Adaptation

For any particular environment, some kinds of organisms survive well, some survive less well, and some cannot survive at all.

The unit focused on how engineering-based criteria can be developed, evaluated, and applied to improve a forest ecosystem. The students recognized the need for more information through focused research and better understanding of science and engineering. Some students recognized the need to weigh criteria, which shaped the final engineered solution. With the teacher's guidance, students also recognized the importance of considering many possible solutions. They realized that their first ideas were not necessarily the best ideas. To come up with a solution, the students needed to gather evidence about different plant species and develop arguments that some species would more likely thrive in the habitat. The vignette illustrates how girls, along with boys, worked together to learn about their woods, to solve a problem by planting native species, and to recognize that they could help restore the environment through engineering practices.

CHAPTER 10

Science and Engineering Practices

Asking Questions and Defining Problems

Define a simple design problem that can be solved through the development of an object, tool, or process and includes several criteria for success and constraints on materials, time, or cost.

Analyzing and Interpreting Data

Analyze and interpret data to make sense of phenomena.

Constructing Explanations and Designing Solutions

Generate and compare multiple solutions to a problem based on how well they meet the criteria and constraints of the problem.

Engaging in Argument From Evidence

Construct an argument with evidence.

The students in the vignette engaged in many science and engineering practices, thereby building a comprehensive understanding of what it means to do science. This section, however, highlights engineering practices as part of performance expectations. *A Framework for K–12 Science Education* derives engineering practices from the steps engineers "engage in as part of their work" (NRC 2012, p. 49).

Engineering practices include Defining Problems. In addition to clearly stating the need to be met or goal to be reached, a problem definition must include a list of criteria or attributes of successful solutions and a list of constraints. The criteria usually include how a solution is expected to function along with which other features are desirable, whereas constraints tend to be the available materials, cost, or time. In this vignette, helping the students shift their understanding of the problem from feeding the animals to providing a long-term source of food was an example of this practice. It was also important for the students to investigate the woods—to research the problem—to better define it.

In *designing solutions* to a problem, it is common for children to jump at the first solution that comes to mind. However, it is often the case that the first solution is not the best. In the vignette, the phenology wheel was especially helpful in helping the students generate a feasible solution. A major teaching point of this unit is about the need to take time to generate a number of solutions and then consider each with respect to the criteria and constraints of the problem.

Crosscutting Concepts

Cause and Effect
Cause-and-effect relationships are routinely identified and used to explain change.

This unit highlights the idea that people can affect the natural environment in both positive and negative ways. In this vignette, the accent was on the positive, as students considered what changes they could make to improve the natural environment to better support the animals that lived there. It is important for the students to recognize that even though they were working with plants and animals, they were "engineering" the changes to solve the problem that they defined.

CCSS CONNECTIONS TO ENGLISH LANGUAGE ARTS AND MATHEMATICS

The *NGSS* include connections to *CCSS* to reinforce the need for integration of English language arts (ELA) and mathematics as part of the science curriculum. The vignette highlights the *CCSS ELA* during science instruction:

- **SL.3.1** *Engage effectively in a range of collaborative discussions (one-on-one, in groups, teacher led) with diverse partners on grade 3 topics and texts, building on others' ideas and expressing their own clearly.*

The *CCSS Math* are also supported in Ms. G.'s science class:

- **3.MD.3** *Represent and interpret data.*

Representation and interpretation of data is essential for accomplishing the task of collecting and applying data in the forest. The students are involved with the application of this standard when they use maps to collect information, create the phenology wheel, interpret the data, and argue for the necessity of engineering design.

EFFECTIVE STRATEGIES FROM RESEARCH LITERATURE

The *NGSS* pave the way for increased exposure to all disciplines of science for all students. This is a breakthrough, in particular, for girls, as research attributes gender disparities in science achievement, college degrees, and careers to an early "experience gap" between girls and boys. The *NGSS* provide an opportunity for teachers to reach girls more effectively because girls perceive a disconnect between school science learning and science career goals (Baker 2013). Research points to three main areas where schools can positively impact girls' achievement, confidence, and affinity with science: (1) instructional strategies, (2) curricula, and (3) classroom and school structure (Baker 2013; Scantlebury and Baker 2007).

First, teachers can use instructional strategies to increase girls' science achievement and to bolster their intentions of continuing on in science. Such strategies include building opportunities to experience phenomena and framing science as inquiry. Girls respond well to strategies that integrate literacy with science. When teachers explicitly focus on metacognitive comprehension strategies by using nonfiction texts in science class, girls' science learning and achievement is enhanced. Girls gain confidence in classrooms where risk-taking is encouraged, teachers support positive messages about girls' competence, and "science is for all" is clearly conveyed.

Second, curricula can improve girls' achievement and confidence in science by promoting images of successful females in science. Schools can enhance girls' engagement in science by adopting curricula that focus on science topics related to the girls' interests. Similarly, girls develop aptitude and confidence toward nontraditional science topics when they are exposed to the topics early. For example, when girls have early exposure, they are more interested in computers and technology. Girls become more motivated toward technology if the curriculum incorporates design and stresses aesthetic aspects of science. In addition, girls respond to topics in physical and biological sciences that they perceive as addressing issues relevant to the real world.

Finally, research supports adjusting classrooms' and schools' organizational structures in ways that benefit girls. For example, after-school clubs, summer camps, and mentoring programs enhance girls' confidence toward science and increase mastery of science content. Girls benefit from science and engineering activities that are intentionally designed to give active roles to all learners. This may occur through thoughtfully planned instructional grouping, pairing girls with friends, and giving every student her own materials to tinker with. Although placing girls in all-girl schools is not known to improve their science achievement, it is sometimes possible to improve results through all-girl groupings within classes containing boys and girls.

Girls and the *Next Generation Science Standards*

FIGURE 10.7.

NGSS AND CCSS FROM VIGNETTE

3-LS4 Biological Evolution: Unity and Diversity

3-5 Engineering Design

Students who demonstrate understanding can:

3-LS4-3. Construct an argument with evidence that in a particular habitat some organisms can survive well, some survive less well, and some cannot survive at all.

3-5-ETS1-1. Define a simple design problem reflecting a need or a want that includes specified criteria for success and constraints on materials, time, or cost.

3-5-ETS1-2. Generate and compare multiple possible solutions to a problem based on how well each is likely to meet the criteria and constraints of the problem.

The performance expectations above were developed using the following elements from the NRC document *A Framework for K–12 Science Education*:

SCIENCE AND ENGINEERING PRACTICES	DISCIPLINARY CORE IDEAS	CROSSCUTTING CONCEPTS
Engaging in Argument From Evidence Engaging in argument from evidence in 3–5 builds on K–2 experiences and progresses to critiquing the scientific explanations or solutions proposed by peers by citing relevant evidence about the natural and designed world(s). • Construct an argument with evidence. **Asking Questions and Defining Problems** Asking questions and defining problems in grades 3–5 builds from grades K–2 experiences and progresses to specifying qualitative relationships. • Define a simple design problem that can be solved through the development of an object, tool or process and includes several criteria for success and constraints on materials, time, or cost. **Constructing Explanations and Designing Solutions** Constructing explanations and designing solutions in 3–5 builds on prior experiences in K–2 and progresses to the use of evidence in constructing multiple explanations and designing multiple solutions. • Generate and compare multiple solutions to a problem based on how well they meet the criteria and constraints of the problem.	**LS4.C: Adaptation** • For any particular environment, some kinds of organisms survive well, some survive less well, and some cannot survive at all. **ETS1.A: Defining Engineering Problems** • Possible solutions to a problem are limited by available materials and resources (constraints). The success of a designed solution is determined by considering the desired features of a solution (criteria). Different proposals for solutions can be compared on the basis of how well each one meets the specified criteria for success or how well each takes the constraints into account. **ETS1.B: Developing Possible Solutions** • Research on a problem should be carried out before beginning to design a solution. Testing a solution involves investigating how well it performs under a range of likely conditions. • At whatever stage, communicating with peers about proposed solutions is an important part of the design process, and shared ideas can lead to improved designs.	**Cause and Effect** • Cause and effect relationships are routinely identified and used to explain change. **Influence of Science, Engineering and Technology on Society and the Natural World. (a)** • People's needs and wants change over time, as do their demands for new and improved technologies • Engineers improve existing technologies or develop new ones to increase their benefits, decrease known risks, and meet societal demands.

CCSS Connections for English Language Arts and Mathematics

SL.3.1 Engage effectively in a range of collaborative discussions (one-on-one, in groups, teacher led) with diverse partners on grade 3 topics and texts, building on others' ideas and expressing their own clearly.

3.MD.3 Represent and interpret data.

CHAPTER 10

CONTEXT

DEMOGRAPHICS

Nationwide, the total number of students in grades 1–12 of public charter and traditional public schools was approximately 49 million in 2009. Slightly less than half of those students, about 24 million, were female (NCES 2011a).

SCIENCE ACHIEVEMENT, ADVANCED DEGREES, AND CAREERS

Test scores in science achievement of female students lag behind those of males. According to the 2009 National Assessment of Educational Progress (NAEP), females received lower scores in science at the three grade levels tested: 4, 8, and 12 (NCES 2011b). Girls were more likely than boys to have insufficient science credits necessary to accomplish mid-level or rigorous curriculum in high school (15% of girls, compared to 9% of boys, had insufficient credits). In addition, even when girls had the same level of class completion as their male counterparts, they received lower scores on the NAEP assessment. Girls who took standard science offerings in high school scored just slightly behind boys in the same classes, while girls taking high school courses at mid- and rigorous levels scored significantly lower than boys taking the same classes.

The research shows that the options for females to pursue science-related fields at the high school, college, and career levels are progressively closed off (NCES 2012). The trend begins in high school, where females take more life science classes such as biology, environmental science, and health science technology and complete fewer advanced classes in physics, engineering science technologies, and computer science. Throughout their high school experience, females opt for fewer mid-level and AP-level courses, and therefore are less likely to meet requirements for an undergraduate science major.

Women have made steady progress since the 1960s in attaining undergraduate degrees, with women nowadays completing undergraduate programs at a higher rate than men (NCES 2011a). However, women are far less likely to complete degrees in technology, science, and engineering fields. In 2010, women made up 58% of two-year college students, but received just 15% of the Associate of Science degrees in engineering technologies (Milgram 2011). In 2006, only 20% of students in physical science and engineering fields were female (NSB 2006). Overall, women are significantly underrepresented in engineering, technical, and computer science careers (NSB 2006).

EDUCATION POLICY

In 1972, Congress passed Title IX, barring gender-based discrimination within federally funded educational programs. Recent policy to change the number of girls going into STEM careers includes the School to Work Opportunities Act, passed by Congress in 1994, to ensure the participation of female students in work transition programs

Girls and the *Next Generation Science Standards*

(School-to-Work Opportunities Act of 1994. P.L. 103-239). More recently, President Obama pushed the "Educate to Innovate" campaign with the goal to "expand STEM education and career opportunities for underrepresented groups including women and girls" (The White House Office of the Press Secretary 2010).

REFERENCES

Baker, D. 2013. What works: Using curriculum and pedagogy to increase girls' interest and participation in science and engineering. *Theory Into Practice* 52 (1): 14–20.

Earth Partnership for Schools Program. 2010. *Earth partnership for schools: K–12 curriculum guide.* Madison, WI: University of Wisconsin.

Forbes, A. 2008. Conceived and developed circular calendar wheel design. Partners in Place. *www.partnersinplace.com/wheels-of-time-and-place*

Milgram, D. 2011. How to recruit women and girls to the science, technology, engineering, and math (STEM) classroom. *Technology and Engineering Teacher* 71 (3): 4–11.

National Center for Education Statistics (NCES). 2011a. *The condition of education 2011 (NCES 2011-033).* Washington, DC: Department of Education.

National Center for Education Statistics (NCES). 2011b. *The nation's report card: Science 2009.* Washington, DC: Department of Education.

National Center for Education Statistics (NCES). 2012. *Digest of education statistics, 2011 (NCES 2012-001).* Washington, DC: Department of Education.

National Research Council (NRC). 2012. *A framework for K–12 science education: Practices, crosscutting concepts, and core ideas.* Washington, DC: National Academies Press.

National Science Board (NSB). 2006. *Science and engineering indicators 2006.* Arlington, VA: National Science Foundation. *www.nsf.gov/statistics/seind06*

Scantlebury, K., and D. Baker. 2007. Gender issues in science education research: Remembering where the difference lies. In *Handbook of research on science education,* ed. S. K. Abell and N. G. Lederman, 257–285. Mahwah, NJ: Lawrence Erlbaum Associates.

The White House, Office of the Press Secretary. 2010. Education knowledge and skills for the jobs of the future. *www.whitehouse.gov/issues/education/k-12/educate-innovate*

CHAPTER 11

STUDENTS IN ALTERNATIVE EDUCATION AND THE *NEXT GENERATION SCIENCE STANDARDS*

Members of the *NGSS* Diversity and Equity Team

ABSTRACT

Alternative education encompasses many nontraditional models, some of which are intended to target students at risk for dropping out. A significant proportion of economically disadvantaged students, racial and ethnic minority students, and English language learners attend dropout prevention schools. State and federal accountability for alternative education has increased, and there is a call to methodically measure the effectiveness of alternative education policy. The *Next Generation Science Standards* (*NGSS*) raise the bar for all students. This magnifies the need for teachers in alternative education to foster engagement and increase exposure to rigorous science. The vignette of a high school chemistry class in an alternative education setting for dropout prevention highlights five strategies: (1) structured after-school opportunities, (2) family outreach, (3) life skills training, (4) safe learning environment, and (5) individualized academic support.

VIGNETTE: CONSTRUCTING EXPLANATIONS ABOUT ENERGY IN CHEMICAL PROCESSES

While the vignette presents real classroom experiences of the *NGSS* implementation with diverse student groups, some considerations should be kept in mind. First, for the purpose of illustration only, the vignette is focused on a limited number of performance expectations. It should not be viewed as showing all instruction necessary to prepare students to fully understand these performance expectations. Neither does it indicate that the performance expectations should be taught one at a time. Second, science instruction should take into account that student understanding builds over time and that some topics or ideas require extended revisiting through the course of a year. Performance expectations will be realized by using coherent connections among disciplinary core ideas, science and engineering practices, and crosscutting concepts within the *NGSS*. Finally, the vignette is intended to illustrate specific contexts. It is not meant to imply that students fit solely into one demographic subgroup, but rather it is intended to illustrate practical strategies to engage all students in the *NGSS*.

INTRODUCTION

Curie Senior High School has a diverse population of more than 700 students. Its motto on the school website attests, "It is never too late to earn a high school diploma." The mission is to deliver a high-quality academic and career/technical program that will lead to a high school diploma or vocational certificate. The school offers traditional and accelerated programs: GED preparation; External Diploma programs; and vocational programs including automotive technology, barbering, cosmetology, Microsoft Office courses, and culinary arts. The median number of years of high school a typical student has attended prior to enrollment at the school is one year. Seventeen percent of the students are of high school age, 45% are 18–24 years old, 14% are 25–29 years old, and the remaining 24% are over 30 years old. Fifty percent of the students, whose neighborhood boundary school was Curie Senior High School, are in-boundary.

ALTERNATIVE EDUCATION CONNECTIONS

Ms. B.'s 10th- and 11th-grade afternoon chemistry class has an average attendance of 17 students, varying in age from 17 to 26. Ms. B. had already encountered a number of the usual challenges developing a supportive classroom community and maintaining high expectations. As usual, the number of students who were registered for the semester course had dropped significantly after a few months due to truancy. Each day a different assortment of the students greeted her. Teaching was further complicated by the fact that many students had uneven or disrupted school careers and thus had significant gaps in their understanding of basic science concepts. The classroom was outfitted with an interactive whiteboard, black lab tables, and 10 large desks with computer workstations in the corners of the room. The walls were covered with science and engineering posters, the periodic table, student-created historical timeline of the periodic table, and student-constructed chemistry family trees. Class sessions were one hour, 40-minute blocks.

The vignette highlights a public alternative school focused on increasing graduation rates for students considered at risk of dropping out of high school. Throughout the vignette, classroom strategies that are particularly effective for students in alternative education are highlighted in parentheses.

Introducing Career Connections to Chemistry

One of the main interests of Ms. B.'s students was to explore career choices. To this end, prior to the introduction of chemical reactions, Ms. B. and two of her math colleagues took a combined math and science class to a STEM Career Workshop. (*Focusing on career connections is one of the life skills strategies promoted in alternative education.*) The school collaborated with the Central Office of the District to coordinate field experiences in conjunction

Students in Alternative Education and the *Next Generation Science Standards*

with the Science and Engineering Festival. A group of approximately 25 students rode on a chartered bus to the Learning Center downtown. The students were welcomed, registered, and given a choice of workshops to attend.

At the Forensic Science workshop, three students, Deshawn, Rosalee, and David, examined different objects; the office had been transformed into a crime scene. A shoe with a huge footprint was displayed in one corner of the room. Other items had been placed in the office. The students drew an outline of the office and the shapes of the objects they encountered in their notebooks. The students noticed a white powder on the shoe. Rosalee listed the physical properties for the white powder found on the shoe. She looked at the powder under a microscope, and noted "tiny cubes, different sizes. Some have knocked off corners with straight sides."

Students from various schools sat in their chairs and went over their observations and the evidence they collected. Deshawn predicted that the unknown white compound was salt and Ms. B. asked how she had come to that conclusion. Deshawn replied that she remembered the introductory lab on physical and chemical changes. Rosalee described the substance in terms of color, odor, and texture. She too thought it was salt, and noted that the white solid dissolved in water and made the temperature of the water go down slightly, indicating a chemical reaction. Ms. B. agreed that their findings were consistent with it being a salt, but added that they would need to do some additional investigating to test if it was a salt and what kind of salt it was. Ms. B. asked Rosalee to predict what elements were in the chemical compound she observed for salt. Rosalee took out her notebook and looked at her periodic table. She then wrote sodium and chlorine on a piece of paper. Ms. B. noted that Rosalee identified her predicted substance as sodium chloride.

The facilitator explained that forensic scientists find, examine, and evaluate evidence in a crime scene. The facilitator asked students what skills might be good for someone entering a forensic science career. One student said good observation skills, another said good reasoning skills or logic, and still another said chemistry. As Rosalee came away from the forensic science workshop, she remarked to Ms. B. that she had learned that forensic science involves chemical reactions.

The next day, Rosalee and Deshawn collected evidence for the conclusion they had reached the day before. They studied some compounds to see if they could predict chemical reactions of ionic compounds on their own. Rosalee took out an interactive science notebook and reviewed her article on salt. Her task was to find the author's central idea of the passage and locate evidence that supported her prediction of the compound's chemical name, sodium chloride. Rosalee explained to Deshawn the properties she discovered about salt: dissolves in water, cubic-shaped, crystalline, white color, a compound with ionic bonds forming from a metal and a non-metal (DCI: HS.PS1 A Matter and Its Interactions). She also described the main idea of the passage and the evidence she thought supported the conclusion that salt, sodium chloride, was on the shoe in the crime lab (practice: Obtaining,

CHAPTER 11

Evaluating, and Communicating Information). Rosalee and Deshawn recorded the chemical formula of salt accurately.

Introduction to the Core Idea: Finding Patterns in the Periodic Table

Ms. B. wrote the driving question for the next few weeks in large letters on the board: "Why do some substances react and others don't?" This question would serve as the focus for questions and discussions, guiding the students' written reflections in their journals (practice: Asking Questions and Defining Problems). One of the subquestions they explored was the energy changes that might take place when ionic structures form and dissolve (DCI: HS.PS1.B Chemical Reactions). On a video that they watched in class, the explosive reaction of sodium metal to chlorine gas formed sodium chloride. The class developed a claim that energy changes due to electrical interactions, by building on explanations for why various materials react. They formed partners and collaborated on their reflections using their science notes, the periodic table, and their initial understandings of elements. (*Student mentoring is an academic support strategy that promotes engagement.*)

Earlier in the month, while studying static electrical charges, students constructed the explanation that positive ions attract negative ions, positive ions repel positive ions, and negative ions repel negative ions. Their explanations were supported by testing potential compounds of metals and nonmetals on element cards with positive and negative superscripts on the upper right side of the cards. The class had turned the cards into a matching game, gathered testable scenarios in their notebooks, and attempted to find patterns.

David, Deshawn, and Rosalee, along with their classmates, applied their organization of student-created chemistry family tree models to the periodic table model and divided the elements into groups and periods. They had developed a beginning conceptual understanding of the patterns of different behaviors of elements on the periodic table: electronegativity, ionization energy, and electron affinity. Conceptualizing chemical properties as patterns helped the students build an understanding of the core idea (DCI: HS.PS1.A Matter and Its Interactions; CC: Patterns). Provided with two small whiteboards, students diligently worked together to draw models showing the electrical attraction involved among atoms when forming ionic compounds. They also used the models to predict what would happen when various ionic compounds were dissolved in water (practice: Developing and Using Models).

Ms. B. used an analogy of a tough rubber band to describe the energy involved in an endothermic reaction that they were familiar with: "The ions in the sodium chloride salt have to be pulled apart. The force that holds them together is like a rubber band. If we give them energy to move apart, the rubber band will stretch. To pull the ions away from each other requires energy, similar to stretching the rubber band." The students played around with big rubber bands for a while and wrote down a corresponding rule in their notebooks: "As ions get pulled farther apart, they are taking a lot of energy and cooling everything down. Breaking bonds is endothermic." Ms. B. told them that they would need this idea to understand the exothermic and endothermic processes they would be working on throughout the following week (DCI: HS.PS1.B Chemical Reactions).

Ms. B. told the class, "Water is also involved in these energy interactions, and I want you to observe these magnets." The students pulled magnetized balls apart to place a marble in the middle of the balls. They discussed when the effort had taken up energy, and when the task had "given up" energy. Ms. B. asked if they thought that, in the crime scene experience with sodium chloride and H_2O, either compound was getting pulled apart. She asked, "Are you releasing energy to the system or taking energy when you pull the magnets apart?" She added that water molecules have both positive and negative ends that pull the water molecules together similar in effect to magnetic forces pulling magnets together.

Developing Explanations for Chemical Properties of Matter

David, Deshawn, and Rosalee had decided to stay after school to work on the experiment on chemical interactions. At 4 p.m., as the other students slowly filed out, Ms. B. bustled around the small lab, putting things away and cleaning up. She nodded to Deshawn and said she would be with them in a minute. (*Providing structured after-school opportunities is an effective strategy for alternative education students.*)

David, Deshawn, and Rosalee were building on their previous learning by exploring endothermic and exothermic chemical interactions. To engage with the crosscutting concept Energy and Matter, the students considered the question, "How is the temperature of water affected when calcium chloride is mixed in it?" David and Rosalee joined Deshawn at the lab table. The partners were relaxed and easy with each other and with Ms. B. They enjoyed spending time in the science classroom and appreciated Ms. B.'s enthusiasm for chemistry. (*Promoting students' safety is important in alternative education.*)

With her goggles on at a black lab table, Rosalee read over the question, "How is temperature of water affected when materials dissolve in it?" Deshawn read the subquestions: "Is H_2O and $CaCl_2$ an endothermic or exothermic reaction? Or how is the temperature of water affected when calcium chloride is mixed in it?" David adjusted his goggles and looked over the materials in the center of the table: a graduated cylinder filled with about 100 ml of water, a pack of hand warmers, a thermometer, a container of calcium chloride, a container of ammonium nitrate, a container of baking soda, small sandwich bags, measuring cups, and a small sandwich bag filled with iron filings.

"I don't know how the temperature of the water will be affected when combined with calcium chloride," Rosalee announced to her partners. She looked over at Deshawn, who said, "You have to state a claim first. See here." Deshawn pointed to the word "claim" in bold print and read the words, "In your science notebook, write a claim supported by evidence to address this question." David read the question again to himself. Rosalee pulled out a separate sheet of paper to jot down her thoughts. She began to read out loud: "Part 2—How is the temperature of water affected when calcium chloride is mixed in it?"

CHAPTER 11

All three students pondered the question, unsure where to start. After a few minutes of silence and pencil tapping, Ms. B. interjected, "Think about what we did yesterday when we pulled apart the magnetic compounds, and think about whether the process may require energy or whether it will release energy."

Deshawn pointed at Ms. B. with her pencil, "The calcium chloride … it will heat up!" Rosalee started to form her statement, "The temperature … if I combine the water with …" and then she slowly formed the claim on the sheet of paper and transferred the sentence to her science notebook. Rosalee was focused and again recited her statement out loud, "If I mix water with calcium. …" Deshawn helped out, "With calcium chloride, girl."

"Where do you write this at?" David inquired while Rosalee continued forming her thoughts out loud to Deshawn. "The temperature. …" She stopped, noticing Deshawn had something to say. Deshawn said, "I think it's going to get the water warm." David located the little space between the question and claim where he could write the group statement.

Ms. B. acted as a facilitator for the session. "So, when you are making a claim, what are you guys expecting to observe? What are you expecting to look at? I heard Deshawn make a prediction. Why did you predict that the water will get warm when we add calcium chloride, Deshawn?" "You mean calcium like … milk … calcium?" David asked. Ms. B. responded to David, "Yes. This time it is not going to be just calcium; it is going to be calcium chloride, the compound." She pointedly looked at the definitions on the whiteboard that the class had made about ions and ionic compounds. She asked, "Deshawn, what do we know about calcium chloride?" Deshawn responded, "Chloride is in the form of chlorine, but now that it is combined with calcium, it has formed a product." David listened to his partner.

After a few more minutes, Rosalee sighed, "I don't know, Deshawn. I think it is going to be endothermic. The temperature is going to decrease." Rosalee slowly nodded, satisfied with her statement. "The water molecules are going to have to pull apart and move around on their own. That takes energy, like those rubber bands, making everything cool down" (practice: Developing and Using Models).

"What is going to decrease?" Ms. B. asked. Rosalee responded, unsure, "The temperature of the mixture. The calcium chloride and water mixture." Ms. B. inquired, "What is happening with the energy when there is a decrease in temperature?" Deshawn and Rosalee said, "Makes it colder." Deshawn added, "Endothermic, takes energy to pull apart the bonds."

"What questions do you have?" Ms. B. asked the group. Rosalee surmised, "If I combine water with calcium chloride, will the temperature decrease?" Deshawn suggested, "I'll write 'both decrease or increase' 'cause we don't know which it is going to be yet." David asked, "Hold on, at room temperature would you say that water is cold or warm?" "Warm!" Rosalee and Deshawn agreed in unison. "Kinda both," Deshawn offered. Rosalee disagreed: "I say warm."

Students in Alternative Education and the *Next Generation Science Standards*

After another minute of discussion, everyone agreed to the water being warm at room temperature. Rosalee took the temperature and inserted the thermometer into the 100 ml graduated cylinder filled with water. Rosalee measured the temperature and reported the temperature was 16 degrees Celsius. Ms. B. encouraged the group to continue with the procedures using their agreed-upon lab roles from the previous day (practice: Planning and Carrying Out Investigations).

Deshawn read the first step of the procedure. Rosalee repeated her statement, "If I combine water with calcium chloride, the temperature will decrease, be endothermic, and the substance will become a powder, form a powdery liquid." Deshawn said, "No, I would say a solution." Rosalee restated, "If I combine water with calcium chloride, the temperature will decrease, be endothermic, and the substance will form a solution because the calcium chloride mixed in. This is my claim for part 2." Deshawn and David wrote down claims in their own words.

FIGURE 11.1.

STUDENTS WORKING AFTER SCHOOL

Rosalee read the next step out loud: "Pour 10 ml of water in an empty plastic bag." She opened the plastic bag, asking, "Who wants to pour the water into the bag?" Deshawn laughed, "You are irritating." Rosalee grabbed the small measuring cup and handed it over to Deshawn. Deshawn took the graduated cylinder and poured 10 ml of water into her measuring cup (Figure 11.1). She checked her tick marks on her cup. David was eager for Deshawn to place the water into the bag. "You are making me nervous, Rosalee," David said. Rosalee expected Deshawn to hand over the measuring cup, but Deshawn poured the water into the sandwich bag proudly and said, "It is 10 ml. I made sure."

"Feel the bag and observe how the temperature feels and record the temperature in your science notebook," David read. Rosalee directed Deshawn, "Feel that. It's cold." Rosalee observed, "Yeah, it's real cold." Deshawn agreed. David wrote down their observations. Deshawn wrote, "The water is cold. The temperature is cold."

Rosalee read, "Place the thermometer in the bag. Make sure the bulb of the thermometer is in the water. Record the temperature of the water in your science notebook. Leave the thermometer in the bag." She concluded, "When I put the water in the cylinder, it was 16 degrees. When you put the thermometer in the water, it was 17. Now it is going up to 18 degrees. Maybe this water is warmer than that water." The group recorded the temperature of the water in the bag at 18 degrees Celsius.

CHAPTER 11

Rosalee read, "Carefully pour 4 ml calcium chloride into the water and gently mix." Rosalee located a new measuring cup and the calcium chloride. Deshawn opened the sandwich bag filled with iron. She thought it was calcium chloride. "Deshawn, are you sure that what you have is calcium chloride?" Ms. B. warned. Suddenly, there was confusion as to what was on the table. David started to pick other items up, asking "What is this?" Ms. B. encouraged them to check all of the materials on the table.

Rosalee took her textbook and examined the periodic table. Ms. B. asked, "What are the elements for calcium chloride?" Deshawn responded, "Ca and Cl." David found the calcium chloride. Deshawn remarked how similar it appeared to the baking soda. David laughed, "Baking soda is not on the material list." Rosalee grabbed the container of calcium chloride and handed it to Deshawn.

As Rosalee read the instructions out loud, Deshawn poured 4 ml of calcium chloride into the measuring cup and then into the sandwich bag. They tried several times before finally achieving success. "Gently pour," Deshawn murmured as the solid white substance fell out. "It says gently pour. It is coming out. Listen to directions," Deshawn reminded the group.

Rosalee continued reading instructions, "Wait 30 seconds and record the temperature again. Remove the thermometer from the bag and carefully zip the bag. Feel the bag again and note the temperature change. Gently mix. Are you going to mix?" Rosalee tried to use the thermometer to mix the reactants. "The temperature is 44 degrees!" she called as she removed the thermometer.

Deshawn felt the bag. "That thing is hot!"

"It's hot?" Rosalee felt the bag. "I thought it was going to get cold!"

"Aaw! Feel it, David. It got warm." Rosalee laughed.

"Oooh, yes! I knew it!" Deshawn smiled.

Using Evidence to Develop Claims

Ms. B. smiled back, "Was this process of dissolving exothermic or endothermic?" Deshawn paused, "It was exothermic because it got hotter; energy was released." Ms. B. reached for the magnetized balls, "Remember the water molecules? They had to part, right? What about the ions?" After more talking and going over the periodic table and their drawings of the electric structure of the ions in the calcium chloride, the threesome determined that they also had to separate (practice: Developing and Using Models).

"There are things that are coming apart, which, as you predicted, would have caused the temperature to decrease, but there are also things coming together." Ms. B. held up the cluster of magnetized balls around the marble, "Which one was more important here?" With the student's help, Ms. B. listed what was coming apart and what was coming together in the process.

Students in Alternative Education and the *Next Generation Science Standards*

Rosalee reviewed for the group, saying, "The Calcium ions had to be pulled away from the chloride ions, and the water molecules also had to be broken apart." Ms. B. said, "Those both took energy, right? Pulling magnets apart takes energy" (DCI: HS. PS1.A Structure and Properties of Matter). David nodded but looked confused. Ms. B. still had the magnetized balls and asked, "What's happening here with these water molecules that are coming together around the ion, the calcium chloride?" She showed the magnetized balls coming together with a loud snap and said, "When the water surrounded the ions, it released lots of energy." Rosalee said, "The bonds clap-snap together, releasing the energy. That makes it get hotter, exothermic. Sometimes it's one and sometimes it's the other." David added, "Energy is making it warm. It has to go someplace; it can't just disappear." Deshawn summarized her thoughts slowly, saying, "Sometimes the things, the bonds that are coming apart, are more important than the things coming together …" Rosalee added, "… and then it is endothermic" (practice: Analyzing and Interpreting Data).

Ms. B. was pleased but planned to strive for clearer and more scientific language from her students. She was about to start a similar discussion with the ammonium nitrate and water endothermic process. She was certain that soon she would be able to bring the conversation back to ions being formed in a way that makes them attract. Students would then be able to make connections with the concept of energy conservation and flow to help them better understand the transfer of energy within a solution (CC: Energy and Matter).

NGSS CONNECTIONS

A Framework for K–12 Science Education (NRC 2012) focuses on the integration of science and engineering practices, crosscutting concepts, and disciplinary core ideas for comprehensive and rigorous science learning for all students. Students who meet these expectations will have the capacity to think critically about science-related issues. They will also have the knowledge and skills to pursue careers in science or engineering if they choose. Students in alternative education, like all students, benefit from meaningful opportunities where they can authentically engage in the three dimensions of the *NGSS*. The lessons described in this vignette help students build toward a deeper understanding of the core ideas in physical science PS1.A and PS1.B in the high school grade band. See Figure 11.2 (p. 153) for the comprehensive list of *NGSS* and *CCSS* from the vignette.

CHAPTER 11

Performance Expectations

> **HS-PS1-2 Matter and Its Interactions**
>
> Construct and revise an explanation for the outcome of a simple chemical reaction based on the outermost electron states of atoms, trends in the periodic table, and knowledge of the patterns of chemical properties.
>
> **HS-PS1-4 Matter and Its Interactions**
>
> Develop a model to illustrate that the release or absorption of energy from a chemical reaction system depends upon the changes in total bond energy.

The lessons in the vignette ultimately lead to an assessment of student understanding as described in the performance expectations above. These performance expectations blend all three dimensions, and the students who are able to perform the objectives described in the performance expectations exhibit grade-level understanding. Even though the students in this vignette initially required a reinforcement of core ideas from the middle school grade band, the expected learning outcomes are still drawn from the high school grade band. The objectives for alternative education students remain at the rigorous level designated for all students.

Disciplinary Core Ideas

> **PS1.A Structure and Properties of Matter**
>
> The periodic table orders elements horizontally by those with similar chemical properties in columns. The repeating patterns of this table reflect patterns of outer electron states.
>
> Stable forms of matter are those in which the electric and magnetic field energy is minimized. A stable molecule has less energy than the same set of atoms separated; one must provide at least this energy in order to take the molecule apart.
>
> **PS1.B Chemical Reactions**
>
> Chemical processes, their rates, and whether or not energy is stored or released can be understood in terms of the collisions of molecules and the rearrangements of atoms into new molecules, with consequent changes in the sum of all bond energies in the set of molecules that are matched by changes in kinetic energy.
>
> The fact that atoms are conserved, together with knowledge of the chemical properties of the elements involved, can be used to describe and predict chemical reactions.

The *NGSS* identify a basic set of core ideas to be mastered by the time a student completes high school. Due to disrupted schooling, students in alternative education may first need to reach standards from previous years to build toward proficiency in current grade-level standards. The vignette illustrates this point by building on incomplete middle school ideas before attempting the disciplinary core ideas required in high school. Revisiting incomplete ideas helps students build and strengthen connections among core ideas.

CHAPTER 11

Science and Engineering Practices

Developing and Using Models

Develop a model based on evidence to illustrate the relationships between systems or between components of a system.

Planning and Carrying Out Investigations

Plan an investigation or test a design individually and collaboratively to produce data to serve as the basis for evidence as part of building and revising models, supporting explanations for phenomena, or testing solutions to problems.

Constructing Explanations and Designing Solutions

Construct and revise an explanation based on valid and reliable evidence obtained from a variety of sources (including students' own investigations, models, theories, simulations, peer review) and the assumption that theories and laws that describe the natural world operate today as they did in the past and will continue to do so in the future.

The students in the vignette engaged in many science practices thereby building a comprehensive understanding of what it means to do science. Students incorporated the science practices of Asking Questions and Defining Problems, Planning and Carrying Out an Investigation, Developing and Using Models, Analyzing and Interpreting Data, and Constructing Explanations and Designing Solutions. This entailed constructing and revising claims from evidence obtained from a variety of sources and experiences. Students developed written and oral explanations from the evidence as they analyzed data and evaluated how well the evidence supported the claims.

Students in Alternative Education and the *Next Generation Science Standards*

Crosscutting Concepts

> **Patterns**
>
> Different patterns may be observed at each of the scales at which a system is studied and can provide evidence for causality in explanations of phenomena.
>
> **Energy and Matter**
>
> Changes of energy and matter in a system can be described in terms of energy and matter flows into, out of, and within that system.

The vignette demonstrates the teaching and learning of the crosscutting concepts to reinforce deeper understanding that can be used to make connections to new scientific ideas. The crosscutting concepts of Energy and Matter and Patterns were found throughout the vignette. The students explored the energy changes in chemical reactions and used patterns observed in the periodic table to predict the behavior of atoms.

CCSS CONNECTIONS TO ENGLISH LANGUAGE ARTS AND MATHEMATICS

All students, including students in alternative education, effectively learn scientific ideas and practices in a cross-disciplinary context. The vignette illustrates how students connect scientific ideas and practices with *CCSS ELA*:

- **WHST.9–10.9** *Use evidence from informational texts to support analysis, reflection, and research.*
  Students engaged with scientific texts in written materials and other informational texts in media.

- **RSLHS 11–12.4** *Determine the central ideas or conclusions of a text; summarize complex concepts, processes, or information presented in a text by paraphrasing the text in simpler but still accurate terms.*
  Students engaged in reading informational texts to determine the author's central idea.

- **RSLS 11–12.4** *Determine the meaning of symbols, key terms, and other domain-specific words and phrases as they are used in a specific scientific or technical context.*
  Students determined the meanings of symbols from the periodic table and key terms, such as endothermic and exothermic reactions, to support their claims of the occurrence of scientific phenomena.

CHAPTER 11

As the students collected data and engaged in experimentation, the vignette illustrates connections to *CCSS Mathematics*:

- **SPM.b** *Reason quantitatively and use units to solve problems.*
- **SPM.d** *Make inferences and justify conclusions from sample surveys, experiments, and observational studies.*

EFFECTIVE STRATEGIES FROM RESEARCH LITERATURE

The term "alternative education" encompasses an array of nontraditional school models (Fitzsimons-Hughes et al. 2006) including, but not limited to, charter, magnet, residential, court, Waldorf, and public alternative schools. These schools are routinely categorized as type 1, 2, or 3 (Gable and Bullock 2006). Type 1 schools are identified as "choice" schools, serve a range of students from gifted students to pregnant teenagers, or have a philosophical or magnet focus (e.g., Afrocentric). Type 2 schools are punitive in nature and may be mandated by court. Type 3 schools have a remedial focus. Although type 3 schools are intended to prevent dropout rates, the research is insufficient to either support or refute this goal. Furthermore, there is virtually no research that relates alternative education to science education.

The research literature focuses on type 3 alternative education, especially schoolwide approaches to promote increased attendance and high school graduation. Specific factors, taken collectively, correspond with alienation from school prior to dropping out. Public alternative schools employ strategies to counteract these factors and increase engagement: (1) structured after-school opportunities, (2) family outreach, (3) life skills training, (4) safe learning environment, and (5) individualized academic support (Hammond et al. 2007).

First, after-school opportunities increase success for students in alternative education. Effective programs offer structured and challenging after-school opportunities in a positive environment that increases engagement with the school. After-school experiences are especially important for students at risk of school failure because these programs fill the afternoon "gap time" with constructive social activities.

Second, effective alternative education programs have family engagement to tackle student alienation through outreach to families. These programs connect schooling with families by organizing field trips, picnics, and other informal activities. They also strengthen families by offering classes on communication, parenting, and student academic support.

Third, effective alternative education programs promote students' life skills, including behavior management, communication skills, and career education. Students improve their ability to make positive choices, express interests, and cope with difficult decisions.

Fourth, effective alternative education programs provide students with a safe learning environment. Although all schools strive to ensure safety, alternative education is sometimes more comprehensive in this area, with policies that forbid harassment and take critical action

Students in Alternative Education and the *Next Generation Science Standards*

FIGURE 11.2.

NGSS AND *CCSS* FROM VIGNETTE

HS. Matter and Its Interactions

Students who demonstrate understanding can:

HS-PS1-2. Construct and revise an explanation for the outcome of a simple chemical reaction based on the outermost electron states of atoms, trends in the periodic table, and knowledge of the patterns of chemical properties.

HS-PS1-4. Develop a model to illustrate that the release or absorption of energy from a chemical reaction system depends upon the changes in total bond energy.

The performance expectations above were developed using the following elements from the NRC document *A Framework for K–12 Science Education:*

SCIENCE AND ENGINEERING PRACTICES	DISCIPLINARY CORE IDEAS	CROSSCUTTING CONCEPTS
Developing and Using Models Modeling in 9–12 builds on K–8 and progresses to using, synthesizing, and developing models to predict and show relationships among variables between systems and their components in the natural and designed worlds. • Develop a model based on evidence to illustrate the relationships between systems or between components of a system. **Constructing Explanations and Designing Solutions** Constructing explanations and designing solutions in 9–12 builds on K–8 experiences and progresses to explanations and designs that are supported by multiple and independent student generated sources of evidence consistent with scientific ideas, principles, and theories. • Construct and revise an explanation based on valid and reliable evidence obtained from a variety of sources (including students' own investigations, models, theories, simulations, peer review) and the assumption that theories and laws that describe the natural world operate today as they did in the past and will continue to do so in the future. **Planning and Carrying Out Investigations** Planning and carrying out investigations in 9-12 builds on K-8 experiences and progresses to include investigations that provide evidence for and test conceptual, mathematical, physical, and empirical models. • Plan and conduct an investigation individually and collaboratively to produce data to serve as the basis for evidence, and in the design; decide on types, how much, and accuracy of data needed to produce reliable measurements and consider limitations on the precision of the data (e.g., number of trials, cost, risk, time), and refine the design accordingly.	**PS1.A: Structure and Properties of Matter** • The periodic table orders elements horizontally by the number of protons in the atom's nucleus and places those with similar chemical properties in columns. The repeating patterns of this table reflect patterns of outer electron states. • A stable molecule has less energy than the same set of atoms separated; one must provide at least this energy in order to take the molecule apart. **PS1.B: Chemical Reactions** • Chemical processes, their rates, and whether or not energy is stored or released can be understood in terms of the collisions of molecules and the rearrangements of atoms into new molecules, with consequent changes in the sum of all bond energies in the set of molecules that are matched by changes in kinetic energy. • The fact that atoms are conserved, together with knowledge of the chemical properties of the elements involved, can be used to describe and predict chemical reactions.	**Patterns** • Different patterns may be observed at each of the scales at which a system is studied and can provide evidence for causality in explanations of phenomena. **Energy and Matter** • Changes of energy and matter in a system can be described in terms of energy and matter flows into, out of, and within that system.

CCSS **Connections for English Language Arts and Mathematics**

RSLHS 11–12.4 Determine the central ideas or conclusions of a text; summarize complex concepts, processes, or information presented in a text by paraphrasing the text in simpler but still accurate terms.

RSLS 11–12.4 Determine the meaning of symbols, key terms, and other domain-specific words and phrases as they are used in a specific scientific or technical context.

SPM.b Reason quantitatively and use units to solve problems.

SPM.d Make inferences and justify conclusions from sample surveys, experiments, and observational studies.

NGSS FOR ALL STUDENTS

CHAPTER 11

to ensure compliance, such as removing offenders. In this way, schools become a safe place for students to learn, an expressed need of alternative education students (Quinn et al. 2006).

Finally, individualized academic support is an effective means of engagement for students in alternative education. Academic support includes specially designed instructional techniques that encourage risk taking, participation, and self-efficacy. Individualized instruction focuses on academics and core subjects and address specific learning needs through mentoring, tutoring, and homework.

CONTEXT

DEMOGRAPHICS

A student in alternative education is defined by attendance in nontraditional schools, which can include public, private, or charter schools. Reporting the demographics of students in alternative education is difficult due to wide inconsistencies in state definitions. States may define alternative schools according to inconsistent criteria, such as the number of students enrolled, a statutory definition of "at-risk," compliance with state regulatory definitions of alternative schools, centralized or independent administration, and funding sources. The three types of alternative education defined above occupy various niches in society and address different needs. Accordingly, students in alternative schools are diverse and come from every demographic group.

Recent data specific to students attending dropout prevention schools do not include nonpublic alternative schools or "push-in" schools-within-a-school. This type of alternative education (type 3) is defined as schools that are "designed to address the needs of students that typically cannot be met in regular schools. The students who attend alternative schools and programs are typically at risk of educational failure" (NCES 2012). During 2010–11, slightly over 500,000 students through grade 12 were enrolled in alternative public schools. A significant proportion of these students were economically disadvantaged students, racial and ethnic minority students, and English language learners (NCES 2012).

SCIENCE ACHIEVEMENT

Science achievement for students who are considered at risk for dropping out, in alternative schooling or another form of nontraditional education, has not been specifically identified on National Assessment of Educational Progress (NAEP).

EDUCATION POLICY

The No Child Left Behind (NCLB) Act of 2001 (the reauthorized Elementary and Secondary Education Act [ESEA]) demanded that stakeholders set high academic standards and increased accountability for the nation's students. In 2008, new federal regulations were adopted, requiring states to use more precise means of counting dropouts. With increased accountability at the federal level, there is a call to methodically measure the effectiveness of alternative education policy (National Alternative Education Association, n.d.).

At the state level, the language for alternative education in laws and policies is inconsistent, ranging from very general to detailed explanations (Hammond et al. 2007). Some states with formal legislation have an official definition for alternative education (sometimes referred to as an alternative program or alternative school). Other states have a policy or language in the law that addresses alternative education funding. In general, state-level alternative education policy is often vague, confusing, inconsistent, and at odds with general education policies. A significant amount of alternative education policy is established locally, and state departments of education may have alternative education efforts that are not captured in law or regulation.

Effective policies for all types of alternative education may contain elements that cut across multiple models or approaches. They must confront multiple risk factors; increase community, family, and student partnerships; and respond to local circumstances. Effective policy responds to the needs expressed in the community and is driven by local initiatives (Hammond et al. 2007). Strategic efforts must target exemplary policies that create innovative pathways for disengaged youth to get back on track to a high school diploma.

REFERENCES

Elementary and Secondary Education Act of 1965, Pub. L. No. 89–10, 79 Stat. 27.

Fitzsimons-Hughes, A., P. Baker, A. Criste, J. Huffy, C. P. Link, and M. Roberts. 2006. *Effective practices in meeting the needs of students with emotional and behavioral disorders in alternative settings*. Reston, VA: Council for Children with Behavioral Disorders.

Gable, R., and E. Bullock. 2006. Changing perspectives on alternative education. *Preventing School Failure* 51 (1): 5–10.

Hammond, C., D. Linton, J. Smink, and S. Drew. 2007. *Dropout risk factors and exemplary programs*. Clemson, SC: National Dropout Prevention Center, Communities in Schools, Inc.

National Center for Education Statistics (NCES). 2012. *Higher education: Gaps in access and persistence study*. Washington, DC: U.S. Department of Education, Institute of Educational Sciences.

National Research Council (NRC). 2012. *A framework for K–12 science education: Practices, crosscutting concepts, and core ideas*. Washington, DC: National Academies Press.

Quinn, M. M., J. M. Poirier, S. E. Faller, R. A. Gable, and S. W. Tonelson. 2006. An examination of school climate in effective alternative programs. *Preventing School Failure* 51 (1):11–17.

CHAPTER 12

GIFTED AND TALENTED STUDENTS AND THE *NEXT GENERATION SCIENCE STANDARDS*

Members of the *NGSS* Diversity and Equity Team

ABSTRACT

A precise figure for the number of gifted and talented students in the United States is not available due to the variation in identification processes from state to state. For nondominant student groups, precise figures are further complicated as states typically rely on only one measure, resulting in fewer students receiving gifted and talented education services. These services, furthermore, are uneven across states and in districts within the same state because there is no federal mandate. The lack of national data—at best, limited national data—on science achievement of gifted and talented students makes it even more difficult to address their achievement. Although the *Next Generation Science Standards* (*NGSS*) provide academic rigor for all students, teachers can employ strategies to ensure that gifted and talented students receive instruction that meets their unique needs as science learners. Effective strategies include (1) fast pacing, (2) different levels of challenge (including differentiation of content), (3) opportunities for self-direction, and (4) strategic grouping. The vignette below highlights effective strategies that promote learning life science for gifted and talented students in an inclusive elementary science classroom.

VIGNETTE: CONSTRUCTING ARGUMENTS ABOUT THE INTERACTION OF STRUCTURE AND FUNCTION IN PLANTS AND ANIMALS

While the vignette presents real classroom experiences of the *NGSS* implementation with diverse student groups, some considerations should be kept in mind. First, for the purpose of illustration only, the vignette is focused on a limited number of performance expectations. It should not be viewed as showing all instruction necessary to prepare students to fully understand these performance expectations. Neither does it indicate that the performance expectations should be taught one at a time. Second, science instruction should take into account that student understanding builds over time and that some topics or ideas require extended revisiting through the course of a year. Performance expectations

CHAPTER 12

will be realized by using coherent connections among disciplinary core ideas, science and engineering practices, and crosscutting concepts within the *NGSS*. Finally, the vignette is intended to illustrate specific contexts. It is not meant to imply that students fit solely into one demographic subgroup, but rather it is intended to illustrate practical strategies to engage all students in the *NGSS*.

INTRODUCTION

Park West Elementary School, in a suburb of a major metropolitan city, has about 450 students in grades preK–4. The demographics are 66% white, 4.4% African American, 6.7% Hispanic, 13.5% Asian, and the remainder multiracial. Twelve percent of the students are low-income. Approximately 12% of the students are identified as academically talented and receive accelerated instruction in reading, language, and/or mathematics in a pullout program for grades 3–4. Otherwise, the students remain with their grade-level group for other subjects, including science. The school typically scores in the 90% range on the state assessments.

Mrs. J., the classroom teacher, generally has between 22 and 25 fourth-grade students in a class that reflects the school's demographics. She enjoys teaching science and is always alert to student needs, especially in science. The challenge of teaching gifted and talented students within the diverse classroom is in assessing their background knowledge and then providing productive and engaging additions to the general science curriculum within the time constraints of the daily schedule. After preassessing her students' current understanding either with a pretest or questioning, Mrs. J. compacts the core content to give her students credit for what they have already mastered, so that the gifted and talented students can do independent or enrichment study while the class is engaged with the content that the gifted and talented students have already mastered. Mrs. J. is flexible with her students, allowing a faster pace for the required activities and offering her students extra time to extend their learning.

This vignette showcases the teacher presenting alternative activities that incorporate an increased level of complexity and abstraction and reflect the interests of gifted and talented students. They are academically engaged through the strategic grouping of students with similar interests, and intellectually challenged through the introduction of advanced ideas and student-generated information. In addition, the gifted and talented develop their own goals and evaluate their own work. They form part of the community of learners and reinforce a prevailing quest for scientific understanding that predominates in the classroom. As a result, all students, including Jerry, Allie, Kate, and Bob (identified as gifted and talented within the vignette), have access for entry at a variety of levels of disciplinary core ideas, science and engineering practices, and crosscutting concepts presented in the *NGSS*. Classroom strategies that are particularly effective for gifted and talented students according to the research literature are highlighted in parentheses.

Gifted and Talented Students and the *Next Generation Science Standards*

GIFTED AND TALENTED CONNECTIONS

Jerry's naturalist interests were immediately apparent in the first week of school. He chose a classroom book on insects as an independent reading choice and excitedly shared the contents with classmates and Mrs. J. As the teacher preassessed Jerry by questioning him, the depth of his knowledge was evident. Jerry had an extensive knowledge about butterflies in particular, knowing their structures, species, and survival needs. Whenever a "bug" arrived at school, Jerry either knew the name or looked it up to inform the class. He would find the other animals of the food web that interacted with the animal. Jerry was the identification expert. His enthusiasm was contagious. He advocated for planting milkweed and provided seeds to the class so classmates could add milkweed to their gardens to help provide food for monarch larvae. Jerry seemed to possess an unusually high interest in nature, and with the preassessment information, Mrs. J. could enhance the curriculum with more rigorous expectations for Jerry. (*Extending level of content by compacting areas already mastered is differentiation strategy of pacing for gifted and talented students.*)

The topic of study in the fourth-grade science curriculum at Park West was native barn owls' internal and external structures and their functions. Models of owls and their anatomy covered the science bulletin board. Mrs. J. was guiding the class to construct explanations about the functions of the structures and facilitating argumentation. The class was organizing the observations by structures grouped in systems. They used the concept to focus on systems rather than memorization of individual structures of the owl itself. Owls were used as the means by which the students studied the overall concept (CC: Systems and System Functions).

During a formal pretest, Mrs. J. found that Jerry knew most of the content of the unit. She decided that praying mantis insects could form the basis of an interest center, functioning as an anchor activity to extend Jerry's learning. Two live praying mantises were brought into the classroom along with flightless fruit flies, and students named the mantises "Lost" and "Found." Jerry immediately volunteered when the teacher suggested that they needed someone to organize the maintenance of the praying mantises and their food. He found books that included the insects, and a learning center began for those students who went into a deeper study of the animals. (*The teacher promoted autonomy and authentic connections to the content.*)

The learning center became a focus for classroom activity. Not only was Jerry an active participant, but so were several other students, including Allie and Kate. The center was populated with books. Then the students sketched the praying mantises and fruit flies, observing details in the body and researching their organs. The sketched diagrams became intricate models, explaining the different functions of the animals' structures. The teacher met with Jerry and other students to discuss the goal of the interest center and their learning. She reinforced that the goal was not only to describe the structures of the animals, but also to construct an explanation based on evidence about the function of those structures (practice: Constructing Explanations ansd Designing Solutions; DCI: LS.1A Structure and

CHAPTER 12

FIGURE 12.1.

INSECT OBSERVATIONS

FIGURE 12.2.

REPORT TO CLASS ABOUT INSECT OBSERVATIONS

Function). (*The teacher promoted autonomy for her gifted and talented students to follow interests and actively participate in their learning objectives.*)

Differences between the two praying mantises were noted and questions came up as to why they were not identical. Students became scientists as they observed (Figure 12.1), looked up information in books and online, and wrote reports answering questions that were posed by other students or the teacher. The reports were posted on the bulletin board, and new information was constantly added with disorganized sticky notes to suggest modifications and more details that would refine or add to the models (Figure 12.2). Questions about other insect species such as bees and ants led the students to develop arguments about the functions of the structures they observed in the praying mantis, comparing them with other classroom insects (practices: Asking Questions, Engaging in Argument From Evidence; DCI: LS1A Structure and Function).

The students, led by Jerry, took responsibility for the insects and their learning center. The small group that condensed around the center's activities conferred daily about the growth and health of the animals. Jerry was able to find the time for the center activities because of the flexibility of pacing in the regular curriculum. He was able to complete other classroom tasks more quickly than his peers; the flexibility with pacing allowed Jerry and others to continue their investigations.

A problem arose as Jerry became alarmed that the food source for the praying mantises was running out. The group of students brainstormed solutions and voted for the teacher to buy mealworms. This new insect became another subject for observation and documentation. Jerry had collaborators, Allie and Kate, who were also passionate about insects. (*Grouping students of similar interests and ability shows a successful strategy for gifted and talented students.*)

The mealworms went through their life cycle and the small group of students informed the class about their progress. Allie described the changes almost daily (practice: Analyzing and Interpreting Data). Differences between the individual mealworms were documented and discussed. Kate researched the insects and her reports were displayed in the interest center (practice: Obtaining, Evaluating, and Communicating Information). Kate's reports on the internal structures were helpful to the group as the students created models comparing the internal and external structures of the mealworms, praying mantises, and fruit flies. Questions about reproduction were also addressed when small mealworm larva appeared. The group shared their findings with the class, and the teacher assessed the learning partly based on the reports the students created.

FIGURE 12.3.

REPORT TO CLASS ABOUT PLANT GROWTH INVESTIGATION

Another opportunity for independent study arose out of a science lesson about bulbs as a food source for a growing plant. Mrs. J. brought an amaryllis bulb into the classroom for the purpose of the lesson. Bob was a quiet child who worked hard. He had a strong interest in sketching. When the potted amaryllis bulb was placed on the windowsill, Bob asked the teacher if he could document the growth of the plant using stop motion photography (DCI: MS.LS1-7 From Molecules to Organisms). Bob collaborated with Jerry and a new interest center developed. They created a chart and a way to standardize the pictures and investigate the growth of the plant. They decided on the measurement tools and controlled the variable of sampling time (practice: Planning and Carrying Out Investigations).

When Bob and Jerry had collected enough pictures on plant growth, they created and narrated a video to illustrate the rate of growth (Figure 12.3). When the flowers bloomed, the students had questions about pollination. The idea of plant reproductive structures is a middle school topic. (*The incorporation of standards from an advanced grade—higher-level core scientific ideas—is an effective strategy for presenting a higher level of challenge for gifted and talented students*.) The students described the structures responsible for reproduction in the flower and developed scientific arguments about how the flower could be pollinated (practice: Engaging in Argument From Evidence). After discussing pollination methods with the teacher, Jerry decided to use a paintbrush to cross-pollinate the flowers. When the flowers dried, the students were initially disappointed that seeds were not produced as they had predicted. After carefully dissecting the flower, there was a surprising discovery: The pods contained very small seeds. This challenged a conception for the students, because they had an assumption that the seeds would be large, based on the size of the flower. The students presented their findings to the class, describing their conclusions from evidence about the seeds.

Independent study of topics of interest led to real-world connections. The independent study was available to all students in the classroom, but the target group was the gifted and talented students. The classroom's wiki website provided an online place to collaborate and share interests. Students chose a topic of interest, worked alone or invited collaborators, and conferenced with the teacher. Once a topic was chosen, the students narrowed or widened the scope by asking a question to clarify their ideas about the topic (practice: Asking Questions and Defining Problems).

Jerry's report on pit vipers broadened his knowledge of the subject as he found information on the physical characteristics of the animals, answering his question, "What characteristics of pit vipers make them excellent predators?" The wiki provided a versatile location that students could use at school or at home as they developed their interest reports. The science topics from just one year included flying critters, pit vipers, rock classification, falcons, robot characteristics, mastodons, zebra mussels, eclipses, and the end of the Earth. These "interest projects" provided an opportunity for students to develop an expertise on self-selected topics and independent research skills. (*The interest projects combine the effective strategies of fast pacing, differentiated challenging content, and opportunities for self-direction including grouping choices.*)

NGSS CONNECTIONS

The vignette illustrates gifted and talented students in a science classroom with a diverse mix of students. It highlights a range of effective classroom strategies, such as learning centers and interest projects, to support and challenge these students in the regular classroom. Students' work was informally assessed based on the products of their studies and questioning. The teacher addressed the unique learning needs of gifted and talented students who developed an understanding of science according to the three dimensions of the *NGSS*. The teacher designed her instruction to include higher grade band standards at a deeper and more challenging level. See Figure 12.4 (p. 167) for the comprehensive list of *NGSS* and *CCSS* from the vignette.

Gifted and Talented Students and the *Next Generation Science Standards*

Performance Expectations

> ### 4-LS1-1 From Molecules to Organisms: Structures and Processes
>
> Construct an argument that plants and animals have internal and external structures that function to support survival, growth, behavior, and reproduction.
>
> ### MS-LS1-4 From Molecules to Organisms: Structures and Processes
>
> Use argument based on empirical evidence and scientific reasoning to support an explanation for how characteristic animal behaviors and specialized plant structures affect the probability of successful reproduction of animals and plants respectively.

Disciplinary Core Ideas

> ### LS1.A Structure and Function (by the end of grade 5)
>
> Plants and animals have both internal and external structures that serve various functions in growth, survival, behavior, and reproduction.
>
> ### LS1.B Growth, Development, and Reproduction of Organisms (by the end of grade 8)
>
> Animals engage in characteristic behaviors that increase the odds of reproduction.
>
> Plants reproduce in a variety of ways, sometimes depending on animal behavior and specialized features.

The students were engaged in the disciplinary core ideas in life science. They explained the structures of animals and their functions. They also explored the role of specialized plant structures in the reproduction of plants, including the role of specific animal behaviors that lead to successful plant reproduction.

CHAPTER 12

Science and Engineering Practices

> **Asking Questions and Defining Problems (by the end of grade 5)**
>
> Ask questions that can be investigated and predict reasonable outcomes based on patterns such as cause and effect relationships.
>
> **Planning and Carrying Out Investigations (by the end of grade 8)**
>
> Conduct an investigation to produce data to serve as the basis for evidence that meet the goals of an investigation.
>
> **Analyzing and Interpreting Data (by the end of grade 5)**
>
> Analyze and interpret data to make sense of phenomena.
>
> **Engaging in Argument From Evidence (by the end of grade 5)**
>
> Construct an argument with evidence, data, and/or a model.
>
> **Obtaining, Evaluating, and Communicating Information (by the end of grade 5)**
>
> Read and comprehend grade-appropriate complex texts and/or other reliable media to summarize and obtain scientific and technical ideas and describe how they are supported by evidence.

As the unit progressed, the students gained abilities in science practices through their exploration of insects and plants by Asking Questions and Defining Problems; Planning and Carrying Out Investigations; Analyzing and Interpreting Data; Engaging in Argument From Evidence; and Obtaining, Evaluating, and Communicating Information. Students argued from the evidence of their observations about the functions of the structures of plants and investigated plant growth and reproduction. The *NGSS* practice of Asking Questions and Defining Problems was illustrated when students engaged in the observations at the learning center and in the independent study.

Gifted and Talented Students and the *Next Generation Science Standards*

Crosscutting Concepts

Systems and System Models (by the end of grade 5)
A system can be described in terms of its components and their interactions.

Students were able to demonstrate their understanding of the crosscutting concept of System and System Models when they observed the structures of the insects and plants and explained how they functioned within the larger system. They generated information orally and in written and digital formats. The observations over time led the students to develop their models of internal and external structures and functions based on the crosscutting concept of System and System Models.

CCSS CONNECTIONS TO ENGLISH LANGUAGE ARTS AND MATHEMATICS

The *NGSS* are committed to the integration of the *Common Core State Standards* for English language arts and mathematics within the content area of science. In the vignette, the teacher inserted reading and writing objectives of the *CCSS ELA* with all students as part of her science curriculum, and differentiated outcomes for her gifted and talented students.

- **SL4.4** *Report on a topic or text ... using appropriate facts and relevant, descriptive details to support main ideas or themes; speak clearly at an understandable pace.* The science practice of *using models* was seamlessly connected to this standard.

- **W.4.2** *Write informative/explanatory texts to examine a topic and convey ideas and information clearly.*
  The teacher was able to raise the bar with her gifted and talented students who were presenting scientific information and explanations not only in a written and oral format, but also in a digital format. As the students developed their online reports, their work connected to the *NGSS* practice of Asking Questions and the *CCSS*.

The vignette also featured the integration of the *CCSS Mathematics*:

- **4.MD.1** *Know relative sizes of measurements within one system of units including km, m, and cm.*

- **4.MD.4** *Make a line plot to display a data set of measurements in fractions of a unit (½, ¼, and so on).*
  Measuring and graphing plant growth highlighted this standard.

CHAPTER 12

EFFECTIVE STRATEGIES FROM RESEARCH LITERATURE

Gifted and talented students may have characteristics such as intense interests, rapid learning, motivation and commitment, curiosity, and questioning skills. While an "integrated theory-driven program characterized by internal consistency from goal setting to service and evaluation" is recommended (Renzulli 2012), often teachers must make curricular decisions and choose instructional strategies that reflect the academic potential of gifted and talented students and target their unique needs as learners. Based on the research literature, teachers can employ effective differentiation strategies to promote science learning of gifted and talented students in these domains: (1) fast pacing, (2) level of challenge (including differentiation of content), (3) opportunities for self-direction, and (4) strategic grouping (Tomlinson 2005).

First, gifted and talented students benefit from faster pacing than their peers. One strategy that involves pacing is called "compacting curriculum," which permits students to pretest out of curriculum already mastered and condense the content partially learned. Flexible pacing with connected extension activities allows sufficient time to explore the areas of study while avoiding redundancies. Teachers can include options for gifted and talented students to impose their own deadlines. Instead of requiring gifted and talented students to do simply more work, teachers provide differentiated instruction through the use of an anchor activity (motivating task).

The second strategy is to promote the level of challenge so it extends the current mastery level of the student through the use of advanced materials and objectives, expectations for idea generation and creativity, complexity of ideas, and open-endedness (Tomlinson 2005). Gifted and talented students benefit from experiences that are connected to the real world. The content should avoid repetitive tasks and be differentiated to encourage expression and foster higher-level and abstract thinking.

The third strategy is to encourage autonomy by allowing the student to follow and cultivate her or his interests and to play a role in her or his own learning trajectory. The teacher helps the student develop strengths and engage in pursuits for which the student has a passion (Tomlinson 2005). In addition, the teacher can incorporate motivating, authentic connections to science content that allow for student-directed goal setting, exploration, and self-evaluation. Choices that reflect the different learning styles of the student should be included (Renzulli 2012).

Finally, effective teachers encourage flexible grouping to enhance academic and socio-emotional development of gifted and talented students, such as classroom grouping that allows for both individual time on projects and opportunities in groups with like-minded peers in terms of ability and interests. Effective grouping can vary between teacher-selected and student-selected to offer a wide range of experiences.

Gifted and Talented Students and the *Next Generation Science Standards*

FIGURE 12.4.

NGSS AND CCSS FROM VIGNETTE

4 From Molecules to Organisms: Structures and Processes
MS From Molecules to Organisms: Structures and Processes

Students who demonstrate understanding can:

4-LS1-1. Construct an argument that plants and animals have internal and external structures that function to support survival, growth, behavior, and reproduction.

MS-LS1-4. Use argument based on empirical evidence and scientific reasoning to support an explanation for how characteristic animal behaviors and specialized plant structures affect the probability of successful reproduction of animals and plants respectively.

The performance expectations above were developed using the following elements from the NRC document *A Framework for K–12 Science Education*:

SCIENCE AND ENGINEERING PRACTICES	DISCIPLINARY CORE IDEAS	CROSSCUTTING CONCEPTS
Engaging in Argument From Evidence Engaging in argument from evidence in 3–5 builds from K–2 experiences and progresses to critiquing the scientific explanations or solutions proposed by peers by citing relevant evidence about the natural and designed world(s). • Construct an argument with evidence, data, and/or a model.	**LS1.A: Structure and Function** • Plants and animals have both internal and external structures that serve various functions in growth, survival, behavior, and reproduction. **LS1.B: Growth and Development of Organisms** • Animals engage in characteristic behaviors that increase the odds of reproduction. • Plants reproduce in a variety of ways, sometimes depending on animal behavior and specialized features.	**Systems and System Models** • A system can be described in terms of its components and their interactions.

CCSS Connections for English Language Arts and Mathematics

SL4.4 Report on a topic or text using appropriate facts and relevant, descriptive details to support main ideas or themes; speak clearly at an understandable pace.
W.4.2 Write informative/explanatory texts to examine a topic and convey ideas and information clearly.
4.MD.1. Know relative sizes of measurement units within one system of units including km, m, cm; kg, g; lb, oz.; l, ml; hr, min, sec.
4.MD.4. Make a line plot to display a data set of measurements in fractions of a unit (½, ¼, and so on).

CONTEXT

Demographics

Reporting the demographics for gifted and talented students is difficult due to wide inconsistencies in the definition, in assessments to identify them, and in funding for gifted and talented programs across the nation.

First, the definition of gifted and talented students varies from state to state and the demographics shift accordingly. Many states have no formal state definition. For this reason it is unrealistic to arrive at an exact number of students in the United States who are gifted and talented. The National Association for Gifted Children (NAGC 2012) defines gifted as, "outstanding levels of aptitude or competence in one or more domains" and estimates that this definition describes approximately 3 million students, roughly 6% of

CHAPTER 12

all K–12 students. There are two more definitions that have been widely applied by states (NAGC 2012). Traditionally, gifted and talented status is for those students performing at the top 5% of an assessment, such as high-stakes testing in language arts or mathematics. An alternative definition is described in Response to Intervention (RTI) as it applies to gifted and talented students. RTI suggests that 5–10% percent of high-performing students in a classroom benefit from strategic, targeted, short-term instruction in addition to core; and 1–6% of the students in a given environment are considered "exceptionally gifted" and need intensive, individualized instruction.

Second, although gifted children come from every demographic group, school districts often rely on only one method of identification. Relying on only one measure may not be effective in identifying the gifted and talented students who come from underserved populations by race or ethnicity, socioeconomic status, and language. In addition, students who have an area of giftedness along with a learning difficulty—referred to as twice-exceptional children—are similarly difficult to identify with only one measure (NAGC 2012).

Finally, as there is no federal funding for gifted programs, school districts must rely on their own funds to support such programs. This results in variations in programming of support for gifted and talented students from state to state. Unfortunately, current policy and funding do not match the needs of students in poverty:

> *One of the barriers to developing the talents of children of poverty is inadequate resources, both financial and in terms of personnel. Developing the talents of any gifted child requires resources for special programs, classes, and support services such as counseling or testing. For children of poverty, even greater amounts of support are needed to help with basic needs of families as well as additional support services such as psychological services for children and families and social workers to assist families with issues surrounding housing and basic subsistence.* (Van Tassel-Baska and Stambaugh 2007)

SCIENCE ACHIEVEMENT

The National Assessment of Educational Progress (NAEP) does not disaggregate for gifted and talented students.

EDUCATION POLICY

In 1988, Congress passed the Jacob Javits Gifted and Talented Students Education Act to fund research grants aimed at better identifying and serving gifted and talented students, especially from underserved student populations. In 2001, ESEA (Elementary and Secondary Education Act, Title V Part D Subpart 6 Dec. 5461-5466) called for a coordination of scientifically based programs to meet the needs of gifted and talented students, and grants to assist agencies and institutions to meet educational needs of these students. ESEA (Title IX, Part A, Section 9101[22]) defines gifted and talented as follows:

> *The term gifted and talented, when used with respect to students, children, or youth, means students, children, or youth who give evidence of high achievement capability in areas such as intellectual, creative, artistic, or leadership capacity, or in specific academic and diversity fields, and who need services or activities not ordinarily provided by the school in order to fully develop those capabilities.*

The above definition has no mandate and only serves as a guide for states that have developed state-mandated definitions. States are not required to use this definition, nor are they federally required to identify gifted and talented students or provide services to them, leaving these decisions to the state and local governments. States, and often districts within states, differ in definitions for gifted and talented students and guidelines for services and teacher accreditation, ranging from full implementation programming to little or none. Although ESEA continued the Javits Act, funds were inconsistent over the years and the act was defunded in 2011 (NAGC 2012).

ESEA calls for the support of state and local efforts to increase the number and diversity of students who participate in and are successful in Advanced Placement courses (Title 1 Part G Sec 1702-1702 cited as the "Access to High Standards Act"). The U.S. Department of Education provides awards to support activities that increase the participation of low-income students in both pre-AP and AP courses and tests.

REFERENCES

Jacob K. Javits Gifted and Talented Students Education Act of 2001. 2001. Elementary and Secondary Education Act, Title IX, General Provisions, 22.

National Association for Gifted Children (NAGC). 2012. *Gifted: Big picture.* Washington, DC: U.S. Department of Education.

Renzulli, J. S. 2012. Reexamining the role of gifted education and talent development for the 21st century: A four-part theoretical approach. *Gifted Child Quarterly* 56 (3): 150–159.

Tomlinson, C. A. 2005. Quality curriculum and instruction for highly able students. *Theory Into Practice* 44: 160–166.

U.S. Department of Education, Advanced Placement Incentive Program Grants. *www2.ed.gov/programs/apincent/index.html*

Van Tassel-Baska, J., and T. Stambaugh, eds. 2007. *Overlooked gems: A national perspective on low-income promising learners.* Washington, DC: National Association for Gifted Children and the Center for Gifted Education, College of William and Mary.

CHAPTER 13
USING THE CASE STUDIES TO INFORM UNIT DESIGN

EMILY MILLER, RITA JANUSZYK, AND OKHEE LEE

Maye, a bilingual Wolof and English speaker, has diligently and painstakingly memorized her sixth-grade science teacher's chart entitled, "Biotic and abiotic factors in an ecosystem." Maye fluently recreates the chart word-for-word on her unit assessment. The teacher walks by her desk, very impressed. He points to the word decomposers *on her chart and asks her, "How do you think the ecosystem dynamics would change if there were fewer decomposers in this system?" Maye thinks for a minute. She can't imagine such an ecosystem. In class they used the chart to discuss the rain forest, a place she had never been. "I think it would change because ... it is all connected and microbes would be less ... less ... decomposing animals and that means less matter." Her teacher nods and then moves on. Maye is relieved to have him move on. She thinks, "I got through that one." And she hopes her answer was the correct one.*

Although Maye could recreate the biotic and abiotic chart for her teacher, she was hard-pressed to apply the science ideas contained in the chart to a new context and demonstrate a robust and flexible scientific understanding referred to as "usable knowledge." Yet, Maye's teacher was following the district curricula to the letter, and he was also meeting state science standards; standards uninformed by the *Framework* and the *NGSS*. The *NGSS* case studies illustrate these documents in action and thus reveal high-leverage shifts that result in units more closely aligned to the *NGSS*.

The unit design in the case studies promotes three-dimensional learning through the following important shifts: (1) the *phenomenon* anchors and frames the unit, (2) the engaging *driving question* guides the direction for learning, and (3) the place-based *context* offers relevance. This chapter will highlight how these three shifts support learning as illustrated in the *NGSS* case studies.

ENGAGING IN PHENOMENA

VIGNETTE FROM A CASE STUDY

The vignette on girls begins in a local woods and wetland in early winter. Having had outdoor experiences throughout the year, the students are familiar with the woods. While

CHAPTER 13

making observations, they take an inventory of available food sources in the winter. Students observe a white-breasted nuthatch walking upside down toward the base of a hickory tree, poking its beak in between the bark crevices. How can this ordinary phenomenon, the white-breasted nuthatch and its gleaning behavior, cause students to wonder about the world? Could this experience of observing the nuthatch's behavior be the avenue to new science understandings? What questions did the nuthatch's behavior raise for the students in the vignette?

The questions vary, bridging crosscutting concepts and science disciplines: *Why is the bird walking upside down and what is it looking for in between the bark? Do other bird species behave similarly? How is the bird's beak structured to catch the food it eats? How does different weather impact the bird's behavior, and what about extreme weather? How many of this kind of bird live in this area? Could you predict whether or not you could find these birds based on the kind of trees in an area?*

The white-breasted nuthatch is an abundant bird species across the United States. After a quick review of research literature on this species, it is apparent that scientists have asked and answered these same questions, and many more. Wondering about phenomena in the natural world is how scientists begin their work. The *Framework* envisions a science classroom that promotes science and engineering, as it is done *authentically* to prepare students with usable knowledge for scientific literacy and college and career readiness. Centering science on engagement in phenomena is part of realizing the vision of the *Framework* and the *NGSS*.

IMPORTANCE FOR THE *NGSS*

The *Framework* enlists students in the task of explaining and predicting phenomena: "The goal of science is to develop a set of coherent and mutually consistent theoretical descriptions of the world that can provide explanations over a wide range of phenomena" (NRC 2012, p. 48). The *Framework* expects even very young students to explain phenomena, as "building progressively more sophisticated explanations of phenomena is central throughout grade K–5, as opposed to focusing only on description in the early grades and leaving explanation to the later grades" (p. 25).

In colloquial terms, people think that a phenomenon is a once-in-a-lifetime extraordinary event, like a comet or a two-headed pig. But the phenomena that interest scientists are repeated and ordinary experiences, which can be explained and predicted. A *phenomenon* is any event in the natural world that happens or will continue to happen under the same conditions.

The *NGSS* transform the way we approach teaching science by reflecting how scientists think and work. When we involve students in phenomena and task them to explain and predict those phenomena, they discover the *usefulness of science ideas* as a means to make sense of the world.

Using the Case Studies to Inform Unit Design

IMPORTANCE FOR DIVERSE STUDENT GROUPS

Science units begin with and are centered on phenomena. This approach reflects an effective practice for diverse students. Students from all backgrounds are immersed in natural phenomena at home and in the community, and they become curious and inspired to ask questions about these phenomena. Teachers need to think critically about how to adapt their units by selecting a commonplace phenomenon that is familiar and engaging to students. Instead of beginning units by teaching abstract science ideas such as erosion, ecosystems, or air pressure, they have students dig pits in the earth, survey a river or woods next to the school, experience rollerblades sliding across a smooth surface, or fill a balloon with air and let it go. They seek out and build from the knowledge that the students have acquired from experiencing these phenomena at home and in the community. Then the students build and refine the ideas of erosion, ecosystems, and air pressure as the means for sense making. The various, rich experiences with phenomena link many different science ideas together so that the ecosystems unit may include ideas about air pressure as well as erosion.

DRIVING QUESTIONS

VIGNETTE FROM A CASE STUDY

The vignette for English language learners (ELLs) revolves around an Earth science unit with the driving question: "Is all soil the same?" This question sparked multiple experiences in the classroom: home interviews, science investigations, and literacy activities. Later in the unit, the driving question was modified to frame the rest of the unit: "Is all of the soil in our neighborhood the same?" and "How can we stop the wind and rain from changing the soil?" These questions sustained the students through the six-week unit, constructing explanations, developing and using soil profile models, collecting and analyzing data, and finally, engaging in engineering design. The unit addressed four performance expectations.

The driving questions mark another key shift in realizing the *NGSS*, as portrayed in the case studies. It is a key shift because (1) it more closely mirrors the work of scientists, and (2) it allows the teacher to place students in the role of sense makers.

What is a driving question? A driving question is the overarching science question that anchors and frames the entire unit. It must be open-ended, engaging, grade-level appropriate, and answerable at various levels of sophistication as knowledge becomes increasingly more specific. With the anchor of the driving question, students see how science experiences are connected and science ideas are related. They gain an enduring understanding about science—science revolves around the process of answering questions.

CHAPTER 13

IMPORTANCE FOR THE *NGSS*

The *Framework* describes the vision for coherence in science by building a K–12 skeletal structure of core ideas with overarching questions that align to each core idea. Although performance expectations change as grades progress, they all refer to these same few overarching questions from the *Framework*. It is these overarching questions that give rise to the driving questions in the units. Similarly, future units that relate to the core idea will address the same driving questions.

For example, the driving questions related to Earth science in the vignette for ELLs are, "Is all soil the same? Is all soil in our neighborhood the same?" and "How can we stop wind and rain from changing the soil?" These driving questions are modified to second-grade terminology from the *Framework*'s driving question for ESS2: "How and why is Earth constantly changing?" Ms. H., the teacher in the vignette, understood that the driving questions in the *Framework* gave rise to the questions her students were involved in, but at an adult level. She adapted the questions to be grade-level appropriate, to bridge different science practices, and to better position her students as sense makers.

IMPORTANCE FOR DIVERSE STUDENT GROUPS

Each vignette in each case study involves students responding to driving questions, not reading and memorizing information. As in the vignette for ELLs, students become motivated to take risks because they sense that the question is authentic, and that there is no simple answer. The driving questions encourage classroom discourse that values different points of view.

Underserved students experience a disconnection between school science and the experiences and understandings about science that stem from home and the community. The disconnection results from the traditional way that science is taught in school, where diverse perspectives are discouraged as irrelevant.

The driving questions are broad enough to spark discussion and argumentation, and they invite contrasting points of view. Through grappling with these broad questions with ever-increasing sophistication, students gain usable knowledge and make sense of the same phenomena in different contexts. When students perceive science questions to have predetermined answers and directionality (e.g., what is erosion?), they are essentially closed off from the benefits derived from soliciting different points of view, perspectives, and experiences. The driving questions underscore the message of inclusion by positioning all students as sense makers.

Using the Case Studies to Inform Unit Design

BUILDING ON PLACE-BASED CONTEXT

VIGNETTE FROM A CASE STUDY

The students in the race and ethnicity vignette learn about the cycling of matter and the flow of energy by comparing and contrasting types of fuel: ethanol, petroleum, and finally, switchgrass. At the time of the unit, switchgrass is a newly developed fuel that is grown and processed in the students' state. In the vignette, the students are encouraged to make connections between science ideas and places they know well: their neighborhood, a local gas station, Nigeria, and Texas. The use of context creates a place for science in the students' lives and emphasizes that science matters and has personal impact. In addition, they become aware of how various fuel sources and the burning of that fuel for energy affect the local economy and the environment in their community.

The students become inspired by the science unit because they see that science and engineering makes a difference in their day-to-day lives. The vignette ends with students writing about how the unit is meaningful to their lives. Novid writes:

> *This is going to affect my future because we use energy every day and we don't want the Earth to get so polluted. Ethanol may be a better solution to keep our Earth clean. Almost every kind of energy has some problems with it; we can still pick one that is the most efficient. Scientists can make better enzymes and protect what we have left.*

IMPORTANCE FOR THE *NGSS*

The *Framework* and the *NGSS* promote science teaching and learning that applies to current societal issues, such as disease prevention, potable water, health care, sustainable energy, and solution of problems related to energy and climate change. The *Framework* highlights that the connection to students' lives is paramount to authentic science learning, and that scientific literacy is a basic tenet of citizenship.

IMPORTANCE FOR DIVERSE STUDENT GROUPS

Context for science is essential to providing access to diverse student groups. The phenomenon that we engage in should be interesting, familiar, and place based (i.e., located in students' local contexts). Place-based science means that science is situated in the local ecology and environment, which encourages students to see themselves as citizens who have a responsibility toward their local community. Science taught in the local context also motivates students to profoundly invest in the science and engineering applications because they understand how this investment directly impacts themselves, family members, and community. In turn, place-based science builds on and values the resources

CHAPTER 13

within the community (e.g., local birders, gardeners, and city planners) and results in partnerships that benefit the school.

Engaging in place-based phenomena allows the students to connect to their cultural and linguistic experiences. One of the main avenues for involving diverse students in science is contextualizing science within areas that are culturally and linguistically meaningful to them. Underserved students, like all students, have experienced a variety of natural phenomena addressed in traditional science units, and they have participated in science practices to make sense of those phenomena at home. For example, most curricula examine weather-related phenomena. Local weather is more likely to drive sense making than otherwise highly engaging weather that the students have never experienced. All students form causal explanations about local weather based on conversations and experiences in their home language. Tapping into these language-based experiences about common experiences gives teachers an optimal place to engage students in making sense of science collaboratively.

APPLICATION OF THE THREE SHIFTS FOR SCIENCE TEACHING WITH DIVERSE STUDENT GROUPS

The catalyst for learning in the vignettes on girls, ELLs, and students from major racial and ethnic groups is the natural *phenomenon*, not the science ideas. Before the students consider the usefulness of science ideas related to the food web, erosion, and energy efficiency, they spend time in the local woods, in the marsh, and at the gas station. Prompted by the *driving questions*, "Do our woods provide enough food for the animals?" "Is all soil the same?" and "What does *efficiency* mean?" they ask questions and make observations about what they encounter. This engagement drives the urgency to talk with one another, do research, collect data, initiate investigations, and solve problems. This resembles the real work of scientists. They immerse themselves in the *context* and compare what they know about a given phenomenon with what they need to know. In college and career, scientific research grows out of a driving question about the phenomenon. Like students in the vignettes, scientists do not merely collect and list examples of animals from different trophic levels, as students might do in a traditional classroom. But rather they walk through the woods with driving questions in mind, observe these organisms in their ecosystem, and develop questions and ideas about relationships between components in the ecosystem to investigate further.

Planning science class to mirror how science is conducted in the real world involves three parts:

1. First, identify an engaging phenomenon that connects multiple science disciplines with various entry points to the grade-level performance expectations.

2. Next, craft one or two driving questions so that they frame the unit and spark discourse through making sense of a phenomenon.

Using the Case Studies to Inform Unit Design

3. Last, ensure that the phenomenon relates to the students' neighborhood and community to engage them to build on and value their prior knowledge. Allow the possibility to address these three steps in the reverse order if there is a pressing and engaging local issue that is relevant to students.

The resulting unit may apply to multiple science disciplines. If there is a stand of trees near an urban school, observe it through the lens of what it could offer students. It can drive overarching questions about Earth materials and water in the hydrosphere (Earth science), different types of matter in the soil and energy from the sun (physical science), the relationship between the birds and insects that live in and around the trees (life science), and a comparison of trees' structures versus manmade materials (engineering design). In an urban setting, even a series of rain events can be a catalyst for sense making, as students can ask questions about the impact of rain on humidity (Earth science), how animals protect themselves in the rain (life science), or the energy contained in moving water (physical science), and the hazards it can cause to people (engineering).

CONCLUSION

The *NGSS* offer new challenges for teachers to reenvision how they will go about teaching science, and how they can approach three-dimensional learning for usable knowledge and comprehensive understanding. Traditional units are based on unfamiliar environments, building seemingly unrelated ideas, and unidirectional questions. In contrast, the *NGSS* are targeted for teaching shifts to engage diverse students. The pressing challenge for teachers of diverse classrooms is to construct a learning environment that provides access to science for students who are underserved and who may not see the purpose of science or view science as relevant to their lives. This chapter demonstrates three key points: (1) The *NGSS* three-dimensional learning can be achieved by highlighting a phenomenon, driving question, and context, which are particularly important for diverse student groups; (2) the case studies provide examples of how to achieve this goal; and (3) teachers can adapt their science units to achieve this goal. This chapter guides teachers to construct science-learning environments informed by the *NGSS* case studies to immerse diverse students in making sense of phenomena.

REFERENCES

National Research Council (NRC). 2012. *A framework for K–12 science education: Practices, crosscutting concepts, and core ideas.* Washington, DC: National Academies Press.

CHAPTER 14

REFLECTING ON INSTRUCTION TO PROMOTE EQUITY AND ALIGNMENT TO THE *NGSS*

EMILY MILLER AND JOE KRAJCIK

What tools are available for teachers to reflect on their teaching both for the alignment to the *Next Generation Science Standards* (*NGSS*) and for the focus on equity? This chapter introduces a rubric for teachers to evaluate their own, and members of their collaborative team's, science teaching. The Equal Access to Language and Science (EquALS) Rubric enables teachers to reflect on their instruction and improve their approach to be more aligned to the three-dimensional learning of the *NGSS*, while paying attention to diversity issues. We use the vignettes in Appendix D "All Standards, All Students" in the *NGSS* (NGSS Lead States 2013) to provide examples of the criteria and their indicators from the EquALS Rubric.

Teachers are striving to transform their instruction based on the shifts recommended by *A Framework for K–12 Science Education: Practices, Crosscutting Concepts, and Core Ideas* (NRC 2012), the *NGSS*, and research-based strategies for diverse student groups. Although several shifts are presented in the *NGSS*, the most significant shift in teaching science, called three-dimensional learning by NRC (2014), combines disciplinary core ideas, crosscutting concepts, and science and engineering practices to enable students in using science ideas, explaining phenomena, and designing solutions to problems. But what does three-dimensional learning look like in science teaching?

THREE-DIMENSIONAL LEARNING

We describe three teaching cases that represent three different approaches to teaching science.

- *Teaching Case 1:* The teacher in this seventh-grade class introduces students to the chemistry unit through a review of the key vocabulary needed for the unit. The teacher has developed many interactive and multimodal experiences to ensure that her students become accustomed to the key vocabulary needed for understanding. She begins with the word wall and asks her students to clap each word and show with one, two, or three fingers their familiarity with each word.

CHAPTER 14

Next, the teacher has the students look up the words and put them into sentences, including a definition and a picture of each word. The students place the words into their science dictionaries. The teacher explains that when the students read and write about the words, they will be prepared to use them appropriately. When the students are tested, they are usually able to match the given vocabulary word to the definition.

- *Teaching Case 2:* Third-grade students begin their Earth materials unit with the idea that rocks are made of multiple substances and that minerals are made of one pure substance. The teacher makes sure that students understand these two ideas through a variety of experiences, including hands-on activities. The teacher has the students take apart a fake rock that is made from a mixture of substances. They describe the substances that make up the fake rock and explain how, together, these substances make up the fake rock. They draw pictures of rocks and minerals and read books that compare the similarities and differences between rocks and minerals. As an assessment, the students explain how rocks are like oatmeal raisin cookies. Many students explain that a cookie is made up of many different things, just like a rock is made up of many different minerals.

- *Teaching Case 3:* Students in a fifth-grade classroom are introduced to the driving question, "Why do changes occur to my body when I exercise?" Students begin the unit outside, running in place and then resting. They compare their active heart rate and respiration rate to their resting rates. Together, the students identify relationships between the variables in the data and develop an explanation for relationships among resting and active heart and respiration rates using diagrams of the different systems in the body. After a few students bring up the possibility of blood circulation and oxygen delivery to the muscles as important, the teacher helps the students develop ideas about cellular respiration and how it occurs. The assessment for the unit involves the students developing a model that accounts for the relationships among the variables in the data they collected. The teacher gives the students a new scenario using insects, observing their respiration, comparing and explaining different respiration rates between insects and humans.

How are the students learning science in the three scenarios above? What are they learning about doing science? Which scenario best aligns with the *NGSS*? And which one is most effective for diverse student groups?

In the first teaching case, the students are learning about science vocabulary, but they are not learning how to apply it in a meaningful way. Although they develop a grasp of the vocabulary related to science by the end of the unit, without the words attached to experiences, they will not be able to apply these ideas. Students learn fragmented pieces of information, but they have no experiences to connect to the words, and they will quickly forget what the words mean. As such, learners will not be able to apply the new vocabulary to new contexts or use it to explain phenomena or to learn new ideas (Bransford, Brown, and Cocking 2000).

In the second teaching case, although students are focused on a scientific idea and have completed some hands-on experiences, the instructional sequence does not anchor the idea in phenomena. The students are expected to restate the idea and apply it to analogies, but they have not generated the idea themselves by experiencing phenomena or engaging in science practices to create explanations. In this scenario, the new science ideas are not connected to experiences of doing science, and as a result many students cannot use the science ideas to make sense of new phenomena. Some students will reproduce the ideas only when prompted with familiar cues.

Teaching Case 3 represents teaching that aligns with the NGSS because it is focused on three-dimensional learning, in which students develop the core idea (transfer of energy and matter), science and engineering practices (developing models, analyzing data), and crosscutting concept (structure and function). The students use three-dimensional learning to explore phenomena and develop useable knowledge that is generative and applied across contexts.

WHAT IS THREE-DIMENSIONAL LEARNING?

What is three-dimensional learning? Why is it valuable? The performance expectations in the NGSS use disciplinary core ideas, science and engineering practices, and crosscutting concepts together to express what students should be able to do at the end of a grade band or grade level. As such, while the performance expectations describe what students should be assessed on, they do not describe how to support students in getting there. For students to meet these performance expectations, they will need to experience the three dimensions working together. When the three dimensions work together, learners will develop knowledge-in-use that will allow them to explain phenomena, design solutions to problems, and build the conceptual structures to learn more. Three-dimensional learning in the NGSS represents a major shift from previous standards. Students can't learn the content without engaging in the science practices, and they can't learn science practices without engaging in the content. Content (disciplinary core ideas and crosscutting concepts) and science and engineering practices work together to help students develop knowledge in-use (Krajcik et al. 2014). Three-dimensional learning is present during the science teaching when (1) each dimension is easily identifiable, (2) the dimensions are developed over the course of a series of lessons that coherently build together, and (3) the working together of the dimensions is not superficial, but essential to carrying out of the series of lessons.

Analogies for three-dimensional learning can be drawn from extracurricular or community-based activities. Take, for example, the way youngsters learn to play basketball in a formal or informal setting. Young basketball players are not taught basketball skills in isolation, nor do they memorize the rules of the game or focus on ideas of sportsmanship. They slowly master how to pass, dribble, and shoot, the rules of basketball, and

CHAPTER 14

sportsmanship—a concept that cuts across all sports—within the context of a real game. Gradually, a coach or other more experienced players lead the youngsters to a more sophisticated understanding of how basketball is played. For example, as the youngsters grow in skill, they are expected to take on increasingly more complex rules as additional skills are brought into the game, such as blocking and free throws. The youngsters are developing the attitude, drive, team play, and interactions with other players that are key to the game. Throughout this process, learners are developing a deep understanding of how to play basketball and taking on an identity as basketball players. Can you imagine how many people would be disinterested in playing basketball if they were forced to learn all the rules associated with the game prior to playing? Or if they had to practice shooting free throws and other skills, devoid of playing a game? Of course, the rules and skills are important; but learning in context is critical to engagement and learning.

In the *NGSS*, students are expected to demonstrate their knowledge of disciplinary core ideas—but the demonstration of knowledge occurs only in the context of doing science. Just like a basketball player, a science student deepens his or her understanding of the natural and designed worlds by applying science and engineering ideas (DCIs and CCs) with science practices to explain phenomena and design solutions to problems. And, just as the basketball player acquires sportsmanship, a science student develops crosscutting concepts inextricably linked with science ideas and practices. The three dimensions work together so learners can build usable knowledge to explain the world around them.

Three-dimensional learning represents an entirely new way of thinking about science teaching and challenges teachers to rethink how they approach teaching science and what it means for students to know science. In the past, practices were used to help students understand the disciplinary core ideas or crosscutting concepts. But three-dimensional learning shows the three dimensions working together to help students make sense of phenomena or design solutions.

HOW IS THREE-DIMENSIONAL LEARNING IMPORTANT FOR DIVERSE STUDENTS?

Anchoring student learning in the three dimensions as described in the *NGSS* reflects current cognitive and pedagogical research about teaching science. Although this approach is challenging for teachers, it is an important shift for all students, and especially critical for diverse students, if they are to develop usable knowledge. The way science is traditionally taught is not working for diverse students. The STEM pipeline has been, and continues to be, essentially cut off to some students, including those in urban and rural poverty settings and English language learners. Devoid of authentic science experiences, students from underserved groups, who may view science as irrelevant to their lives, may not authentically engage with science practices, or they may struggle to connect school science with the science experienced in the home and community.

Reflecting on Instruction to Promote Equity and Alignment to the NGSS

Schools serving underserved student groups can take steps to reverse this trend and transform science teaching. In the three-dimensional learning example in Teaching Case 3, students are actively participating in science (science and engineering practices) as they generate ideas (disciplinary core ideas) to understand how the world is structured (crosscutting concepts). This approach provides for a comprehensive picture of science that underserved students have been denied. The NGSS vignettes offer insights into how to engage diverse student groups in three-dimensional learning.

STEPS TOWARD IMPROVING STEM OPPORTUNITIES FOR UNDERSERVED STUDENTS

The EquALS Rubric will allow teachers to examine their science teaching with respect to alignment with the NGSS. The EquALS Rubric takes into account both the importance of three-dimensional learning and the needs of diverse students. This rubric is designed to observe lessons in action to evaluate if they meet the intent of the NGSS and support all students in learning science. Furthermore, teachers could use the EquALS Rubric to identify areas in their science teaching that need improvement. The rubric includes the following criteria: (1) three-dimensional learning, (2) attention to context, (3) opportunity for discourse, and (4) emphasis on student thinking and reflection (see Appendix A, p. 189). Each criterion of the EquALS Rubric has specific indicators that show the criterion has been met.

CRITERION 1: A FOCUS ON THREE-DIMENSIONAL LEARNING

As discussed above, three-dimensional learning is an essential shift in the *Framework* and the NGSS from previous teaching. If instruction does not meet this criterion, then the teacher needs to reflect on how to modify instruction to have the three dimensions working together.

This criterion is also represented in the Educators Evaluating the Quality of Instructional Products (EQuIP) Rubric developed by Achieve Inc. in collaboration with NSTA (The EQuIP Rubric can be downloaded at *http://nstahosted.org/pdfs/ngss/resources/EQuIPRubricForScienceOctober2014_0.pdf*.) The EQuIP Rubric helps teachers and educators design and evaluate curriculum materials to align with the NGSS: "Elements of the science and engineering practice(s), disciplinary core idea(s), and crosscutting concept(s), blend and work together to support students in three-dimensional learning to make sense of phenomena or to design solutions" (Krajcik 2014).

This focus on three-dimensional learning is equally important to the EquALS Rubric. When examining their instruction, teachers need to take into consideration this critical criterion essential to the enactment of the NGSS. The second indicator of this criterion in the EquALS Rubric states that effective teaching blends the three dimensions in an authentic (nonsuperficial) way so the learning of science reflects the process of doing and thinking science and engineering in real careers. While this criterion is essential for all

CHAPTER 14

students, it is especially important for underserved students. Due to diminished hours in science and fewer opportunities for advanced classes, underserved students are less likely to experience in school how the science practices, crosscutting concepts, and core ideas work together to explain and predict phenomena. Although three-dimensional learning is apparent in all of the vignettes, the vignettes for English language learners, students from economically disadvantaged backgrounds, and students with disabilities provide examples of this first criterion.

- **English language learners.** In the vignette, the three dimensions are easily identifiable. Students use the science and engineering practices of Constructing Explanations and Designing Solutions, Developing Models, Communicating Information, and Solving Engineering Problems. The crosscutting concepts of Cause and Effect and Patterns are reinforced throughout the lessons. The science and engineering practices and the crosscutting concepts are supported within the context of the disciplinary core ideas of ESS2.A: Earth Materials and Systems. Wind and water can change the shape of the land, and ETS1.C: Optimizing the Design Solution. Because there is always more than one possible solution to a problem, it is useful to compare and test designs.

- **Students from economically disadvantaged backgrounds.** Students learn the disciplinary core idea of PS1.A Structures and Properties of Matter concerning the behavior of the molecules in a gas, while working with the science and engineering practices of Developing and Using Models and Engaging in Argumentation From Evidence. They also work with the crosscutting concepts of Energy and Matter and Cause and Effect. In this vignette the students gradually develop the three dimensions over the course of a series of lessons and they coherently build together, starting with an initial model, refining the model over time, and then applying their conceptual model to an engineering problem.

- **Students with disabilities.** In this vignette, the students are engaged in the three dimensions in an authentic way, which supports student engagement and learning about the doing of science. The students *develop and use models* and *analyze and interpret data* to develop the core ideas of ESS1.A The Universe and Its Stars and ESS1.B Earth and the Solar System. This work concerns identifying patterns in the movement of the Sun, Moon, and stars and they also apply the crosscutting concepts of Patterns and Scale, Proportion, and Quantity.

CRITERION 2: ATTENTION TO CONTEXT

Science teaching must be situated in a context that is relevant to students' lives and interests (Krajcik and Shin 2014). The community—rural, urban, and suburban—is a treasury of untapped connections to science and engineering. Teachers who capitalize on context engage students who have become alienated from science. Making use of the resources in the community allows diverse students to bridge their lives to school science. In the

Reflecting on Instruction to Promote Equity and Alignment to the *NGSS*

following vignettes for girls, English language learners, and race and ethnicity, three of the indicators that describe this criterion in the EquALS Rubric are illustrated.

- **Girls.** Students who are engaged in familiar contexts make sense of science by drawing on their linguistic, cultural, and life experiences. In the girls' vignette, students draw from hunting, gardening, and feeding birds to make sense of the engineering design problem in the woods.
- **English language learners.** The context in this vignette shows the teaching that draws from and builds on sociopolitical issues within the community. English language learners become involved with solving a problem of the wind blowing trash into the urban marsh near the apartments where they live. They are inspired to solve the engineering problem not only because it is interesting, but also because they see the benefit to their community. Intentionally connecting science and engineering to local issues can bring in local partnership from the community. Students invest in their neighborhood, and they give the community reason to invest in its children as a reciprocal process.
- **Students from racial and ethnic groups.** The last element of context is connecting science and engineering careers to students' lives. This could be done by bringing in traditional (e.g., chemists, ecologists, industrial engineers) and nontraditional (e.g., gardeners, textile developers and designers, practitioners of traditional ecological knowledge) scientists and engineers from the community or learning about diverse role models who have contributed to science disciplines. In the vignette on race and ethnicity, the students learn from a local expert about the country of Nigeria and the impact the energy industry is having on the environment and economy.

CRITERION 3: OPPORTUNITY FOR DISCOURSE

Science teaching should promote student sense-making through discourse. The focus of teaching is on promoting and validating *all* student ideas by providing supportive structures for soliciting these ideas and leveraging scaffolds to support student participation in science discourse. Promoting opportunity for sense-making through discourse is especially critical for students who are acquiring English as a second language, students with nonstandard English, and students with language processing issues. Connecting science to students' experiences with phenomena, as well as attending to the language needed to participate in science and engineering practices, provides an avenue to use language and engage students in collaboration (Michaels, Shouse, and Schweingruber 2008). The EquALS Rubric lists important indicators for teaching in a way that supports opportunity for discourse, and these indicators are separately illustrated in the following vignettes.

- **Students from economically disadvantaged backgrounds.** Teachers should promote student discourse to drive collective sense-making. As the crux of student

CHAPTER 14

learning resides in student sense-making, the teacher must take a step back to listen to students generating ideas, questioning and challenging, and building on those ideas. In this vignette, Ms. S. divides the students into small groups to develop models of the railroad tank car implosion. The students rely on each other's ideas and questions to construct their model and explain the cause of the tanker implosion. Without this step, or if Ms. S. had led the discussion instead of placing the students in small groups, the opportunity for many students to share their ideas in a safe way might have been overly channeled or truncated.

- **English language learners.** The teacher uses strategies to assist in comprehension and production of language for all students and English language learners in particular. She employs different strategies according to various students' language needs. For example, although all of the students benefit from the jigsaw discourse strategy that places students with certain expertise on different teams so each student is essential and thus compelled to participate, only a few students rely on the pictures that support English words in the assignment.

- **Students in alternative education.** Students participate in discourse with minimal guidance from the teacher, but with intentional cues and prompts. Ms. B. prods the students to check for meaning and to question each other. Only rarely does Ms. B. add her own perspective to the discussion; rather, she serves as a coach who facilitates student participation. Effective teaching models the complex language of science, as well as the *purposes* for language that are inherent to science.

CRITERION 4: EMPHASIS ON STUDENT THINKING AND REFLECTION

Teachers should seek out, value, and build on student thinking and reflection (Michaels, Shouse, and Schweingruber 2008). Teachers make links to, and build from, students' prior knowledge, and also provide the students with the time and tools to reflect on their own learning. Teachers need to use differentiation, alternative assessments, accommodations (including technology), and language supports. Co-construction of science ideas is accomplished in collaborative groups or through establishment of protocols that center on valuing diverse perspectives. Three indicators for this last criterion that emphasize student thinking and reflection are illustrated below.

- **Students from racial and ethnic groups.** Teachers explicitly devise teaching objectives that include metacognitive reflection around three-dimensional learning (Bransford, Brown, and Cocking 2000). As the vignette demonstrates, students need guidance to engage in metacognitive thinking. The students write essays about what they learned throughout the unit and why this learning was important to them. Supporting student reflection about what they did, why they did it, and how their thinking changed as a result of the experience promotes student understanding of three-dimensional learning.

- **Girls.** The vignette on girls emphasizes learning science ideas through various multimodal experiences and solving meaningful problems. As the students gather research about native and nonnative species in the forest, they are able to develop and use the phenology wheel to make their thinking visible, and they see in a concrete way that they are building on each other's ideas. In the end of the vignette, the solution to the problem reflects the entire group's diverse expertise and thought processes.
- **Gifted and talented students.** The vignette shows how students are motivated by teaching that extends or differentiates their learning. For example, students who are interested in the topic of a praying mantis's exoskeleton structure are allowed to extend their interests through independent and partner work (Krajcik and Shin 2014). Other students use technology to promote an interest in digital media and science.

CONCLUSION

Prior to the *NGSS*, science ideas were removed from practices and crosscutting concepts, which resulted in science learning as a memorization exercise that did not enable students to explain related phenomena. At best, inquiry was used in the service of learning science content, rather than the practices with core ideas and crosscutting concepts working together for students to use the knowledge they develop. The major shift represented by the *NGSS* is that all students need to engage in three-dimensional learning to explain phenomena. This is a challenging task, but one that teachers need to embrace to support all students in learning the sophistication of science expected in the *NGSS*.

The *Framework*, the *NGSS*, and Appendix D, along with the vignettes in the seven case studies, provide the route to transform science teaching and learning. Informed by these documents, teachers will begin to reverse the trend toward increasing disparity for diverse student groups by fostering three-dimensional learning, attention to context, opportunities for discourse, and valuing and building on students' strengths and ideas. The EquALS Rubric can help teachers gauge how well their science instruction focuses on these four criteria. The EquALS Rubric can be used to support teachers in reflecting, revising, and refining their instruction to better align with the *NGSS*. The new approach to teaching science that promotes all students' engagement and self-efficacy challenges teachers to rethink how they plan, implement, and reflect on their teaching. They will ultimately reconceptualize their understanding about what it means to teach science and what it means to teach science well.

CHAPTER 14

REFERENCES

Bransford, J. D., A. L. Brown, and R. Cocking. 2000. *How people learn: Brain, mind, experience and school.* Washington, DC: National Academies Press.

Krajcik, J. 2014. How to select and design materials that align to the *Next Generation Science Standards*. NSTA Blog. *http://nstacommunities.org/blog/2014/04/25/equip*

Krajcik, J., S. Codere, C. Dahsah, R. Bayer, and K. Mun. 2014. Planning instruction to meet the intent of the *Next Generation Science Standards*. *The Journal of Science Teacher Education* 25 (2): 157–175.

Krajcik, J. S., and N. Shin. 2014. Project-based learning. In *Cambridge handbook of the learning sciences*, 2nd ed., ed. R. K. Sawyer, 275–297. New York: Cambridge.

Michaels, S., A. Shouse, and H. Schweingruber. 2008. *Ready, set, SCIENCE! Putting research to work in K–8 science classrooms.* Washington, DC: National Academies Press.

National Research Council (NRC). 2012. *A framework for K–12 science education: Practices, crosscutting concepts, and core ideas.* Washington, DC: National Academies Press.

National Research Council (NRC). 2014. *Developing assessments for the Next Generation Science Standards.* Washington, DC: National Academies Press.

NGSS Lead States. 2013. *Next Generation Science Standards: For states, by states.* Washington, DC: National Academies Press. *www.nextgenscience.org/next-generation-science-standards*

Reflecting on Instruction to Promote Equity and Alignment to the *NGSS*

APPENDIX A

EQUAL ACCESS TO LANGUAGE AND SCIENCE (EQUALS) RUBRIC

Criterion 1: A Focus on Three-Dimensional Learning

How well does the science teaching promote three-dimensional learning? How apparent are each of the three dimensions in the teacher's approach?

Indicators: Teaching includes (1) ensuring that the three dimensions are easily identifiable, (2) developing the dimensions over a course of a series of lessons that coherently build together, and (3) the three dimensions working together in an authentic (non-superficial) way, essential to carrying out the lesson or series of lessons.

Level 1	Level 2	Level 3	Level 4
The teaching includes each of the three dimensions; by the three dimensions working together and building over time, the teaching supports the doing of science. The science and engineering practices are used with core ideas and CC so students can explain phenomena or design solutions to problems.	The teaching involves a focus on the three dimensions, and they are each stated as goals, but they do not build together toward understanding. The science and engineering practices are used with core ideas and CC in a superficial manner and lack focus on explaining phenomena or solving problems.	There is potential for the teaching to involve the three dimensions, but they do not build over time, and/or the integration of the dimensions is not observed. The teaching does not support students in doing science and using the three dimensions in an authentic way.	The teaching emphasizes only one or two of the three dimensions in lesson goals. The dimensions included in the lesson are only partially used for building toward understanding. The teaching does not support students in doing science.

Ratings:

Strengths:

Suggestions for improvement:

CHAPTER 14

APPENDIX A (continued)

Criterion 2: Attention to Context
Does the teaching situate science learning in a relevant and meaningful context that builds on home, community, and cultural resources, and engage students to experience phenomena?

Indicators: Teaching purposefully (1) makes explicit connections to student lives and language and culture, involving students' neighborhood and community, (2) engages students in phenomena driven by student-centered questions or problems, (3) uses contexts that students find meaningful to explore, (4) connects science and engineering problems to real-world sociopolitical contexts, (5) uses diverse role models of scientists and engineers, and (6) links science to careers.

Level 1	Level 2	Level 3	Level 4
The teaching offers a strong connection to students' lives and builds on current contexts. The learning hinges on students' experiencing a variety of phenomena and contexts that students find meaningful. Connections to science and engineering in careers are made for students..	The teaching makes some connections to students' lives, but limits or interferes with learning focused on the experiencing of phenomena. Students know the context but don't find it meaningful. The teaching makes occasional reference to science and engineering in careers.	The teaching superficially builds on students' lives and experiences, and connects learning to the phenomenon only indirectly. Students do not find the context meaningful. The teaching may make some reference to science and engineering in careers.	The teaching does not connect students' lives to science, and/or does not include learning based on experiencing of phenomena. Students do not find the context meaningful. There is little or no reference to science and engineering as it is carried out in careers.

Ratings:

Strengths:

Suggestions for improvement:

APPENDIX A (continued)

Criterion 3: Opportunity for Discourse

To what extent does the teaching promote meaningful discourse around sense making and problem solving, and support all students including English language learners, students with difficulties processing language, and nonstandard English speakers in acquiring the language of science? How are speaking, listening, reading, and writing integrated to enhance discourse?

Indicators: Teaching includes (1) meaningful and guided support for language development, (2) support for and expectation of student discourse, (3) explicit models and embedded opportunities to practice the language of science with links to students' home language, (4) clear reliance on overarching ideas, (5) expectation for students to use evidence to support their positions, and (6) scientific and technical vocabulary anchored in phenomena and experiences.

Level 1	Level 2	Level 3	Level 4
The discourse and literacy opportunities are supported and clearly defined, varied, and conceptualized based on making sense of relevant experiences in the classroom and solving problems. Students use evidence to support their claims.	The discourse and literacy opportunities are supported and clearly defined and the purpose of the discourse is present, but authenticity is needed to demonstrate its relevance and importance in the sense making of phenomena. Students only infrequently use evidence to support their claims.	The literacy and discourse opportunities need to be better supported and conceptualized, and has contrived purpose around sense making and problem solving is needed. Students are not required to use evidence to support their claims.	The discourse and literacy opportunities in the lesson are not supported or clearly defined or are missing entirely. Students are not required to use evidence to support their claims.

Ratings:

Strengths:

Suggestions for improvement:

APPENDIX A (continued)

Criterion 4: Emphasis on Student Thinking and Reflection
Does the teaching pay attention to students' current understanding and ideas, use a variety of formative assessment to support student learning, deliver opportunities for differentiation and co-construction of learning, provide scaffolding of challenging tasks and/or extend learning when appropriate?

Indicators: Teaching purposely (1) solicits and builds on current and prior science ideas from the class, (2) uses technology in a meaningful way in authentic contexts, (3) includes time for student metacognitive reflection, (4) engages students in multimodal experiences, (5) offers collaborative groups to participate in tasks or solve problems, and (6) builds on students' worldviews and epistemologies.

Level 1	Level 2	Level 3	Level 4
The teaching is student centered, well-reasoned and accesses and builds on ideas. Teachers and students co-construct understanding. Evidence of building on varied methods of assessment is present. Links to students' prior ideas are made.	Most of the teaching accesses and builds on student strengths and ideas. However, there are aspects of differentiation, social construction of learning, and student-centered instruction that are missing. Some links to prior knowledge are made. Building on assessment data is present, but vague or unvaried.	The teaching attempts to be focused on student ideas and strengths. It is not always clear if the differentiation is based on rationale in terms of student current understanding and student learning outcomes. Assessment does not inform teaching moves. Few links to prior knowledge are made.	The teaching is not organized around student ideas or strength. There is no clear and sound rationale present for the teaching decisions or differentiation. No assessment is included or only a summative assessment is issued. No links to prior knowledge are made.

Ratings:

Strengths:

Suggestions for improvement:

CHAPTER 15

CASE STUDY UTILITY FOR CLASSROOM TEACHING AND PROFESSIONAL DEVELOPMENT

EMILY MILLER, RITA JANUSZYK, AND OKHEE LEE

Meeting the *NGSS* is a daunting prospect for teachers who are already juggling numerous school, state, and district initiatives. The *NGSS* motivate teachers to think about what it means to learn science and how they consider new approaches to science instruction. The *NGSS* case studies offer accounts of real teachers and real students, and their authenticity is meant to be inspirational. We hope that our readers consider these case studies with an open but critical mind to start conversations, projects, and initiatives in science education.

The case studies as a whole or individually can serve as a resource for professional development. Teachers and professional learning communities at the grade, department, school, district, or state level can discuss and analyze the case studies and apply three-dimensional learning and effective strategies to their own contexts.

PROFESSIONAL DEVELOPMENT CONSIDERATIONS

1. *Instruction with three-dimensional learning*
   Consider reading the case studies with your own teaching practice in mind. Use the case studies as illustrations of useful ideas for teaching science. Identifying examples of each of the three dimensions in the vignette can help you as you design instruction in your science class.

   - Think about what ideas, approaches, and strategies from the vignette could inform your instruction. And be critical of the teachers' methods as they are developing their skills to teach for three-dimensional learning combined with effective strategies.
   - While reading the vignette, try to identify each of the three dimensions and explore the ways that they are blended. For example, do you notice a teacher using the crosscutting concept of Structure and Function while supporting students in developing models?

CHAPTER 15

- Do you see students grappling with a disciplinary core idea through the science practice of Asking Questions? What other practices are developed at the same time as the disciplinary core idea?

2. *Instruction with effective strategies*
Each case study focuses on one demographic group of students and explains the classroom strategies that are effective for engaging the group in science. Then, the vignette presents the strategies within a classroom setting of diverse student groups.

    - Read the section on effective strategies near the end of the vignette for the demographic group, and then read the vignette to locate those strategies. You will find that some of the strategies are ones you use already, and some you would like to try with your class.
    - Think about which strategies could be useful for all of your students regardless of demographic groups. Study the vignette for instances of all the students being motivated by the effective strategies.
    - Observe the teacher-student and student-student interactions. How does the teacher address or respond to student ideas? How do the students respond to one another's ideas? What classroom procedures are needed to create this type of classroom culture?

3. *Science achievement data and education policy*
Each case study provides an overview of science achievement data and education policy. Each student comes to the classroom with intellectual resources from home and community, which teachers can draw from when teaching science. We encourage our readers to approach the case studies from this perspective, especially in light of science achievement data and education policy.

    - The case studies provide examples of diverse student groups succeeding when presented with STEM learning opportunities. How can this information impact decisions at the school, district, or state level?
    - Can you find discrepancies between the messages about the capabilities of students in the vignette and the expectations and opportunities offered to students in your school or district?

Most of all, the case studies nudge the conversation to center on *science and diversity* in a productive, constructive way. Every teacher who loves to teach science experiences how science learning inspires all students. Science can motivate even the most reluctant learners, and teachers notice that *all students* bring ingenuity to science class! These professional development considerations that apply to all the case studies as well as the Reflection Guides (Tables 15.1–15.8, pp. 195–202) in this chapter provide direction for the utility of the case studies.

Case Study Utility for Classroom Teaching and Professional Development

TABLE 15.1.
GENERAL REFLECTION GUIDE

Reflection Guide for Case Studies		
Student group:		
Three-dimensional learning	Effective strategies	Science achievement and education policy
Guiding questions		
How did the strategy employed in the vignette create engagement of the focus group?		
By applying three-dimensional learning and effective strategies to your setting, what implications do you see? What examples can you provide from your classroom or school?		
What questions does the vignette raise?		

CHAPTER 15

TABLE 15.2.
REFLECTION GUIDE FOR ECONOMICALLY DISADVANTAGED STUDENTS

Reflection Guide for Economically Disadvantaged Students		
Three-dimensional learning	Effective strategies	Science achievement and education policy
Guiding questions		
How did the strategy employed in the vignette create engagement of the focus group?		
By applying three-dimensional learning and effective strategies to your setting, what implications do you see? What examples can you provide from your classroom or school?		
What questions does the vignette raise?		

1. The teacher had the students develop and use different models. Can you trace the progression toward mastery of this practice throughout the vignette? How did the teacher support each new opportunity to develop and use a model?

2. Can you analyze how the teacher assessed the students' prior knowledge of the disciplinary core ideas and then designed her lessons to build on that knowledge? What would have happened if she had not assessed their prior knowledge?

3. How did the crosscutting concept of Cause and Effect contribute to student understanding of the disciplinary core ideas and/or science and engineering practices?

4. What strategies were particularly effective with economically disadvantaged students? What evidence from the vignette led you to see them as effective? Was there any strategy from the vignette that resonated with you?

Case Study Utility for Classroom Teaching and Professional Development

TABLE 15.3.
REFLECTION CHART FOR MAJOR RACIAL AND ETHNIC GROUPS

Reflection Guide for Major Racial and Ethnic Groups		
Three-dimensional learning	Effective strategies	Science achievement and education policy
Guiding questions		
How did the strategy employed in the vignette create engagement of the focus group?		
By applying three-dimensional learning and effective strategies to your setting, what implications do you see? What examples can you provide from your classroom or school?		
What questions does the vignette raise?		

1. How did the teacher support the students in the science practice of Engaging in Argumeent From Evidence? Did this practice occur spontaneously or did the teacher plan it? What evidence from the vignette leads you to think that?

2. What disciplinary core ideas were included in the vignette? Were there any disciplinary core ideas that needed to be repeated or reinforced? How did the teacher promote student understanding that energy and matter are different, not the same?

3. How did the crosscutting concept of Energy and Matter contribute to student understanding of the disciplinary core ideas and science and engineering practices?

4. What strategies did you find most effective for students from diverse racial and ethnic backgrounds? What evidence from the vignette leads you to see these strategies as effective? What strategy used in the vignette made you reflect on your own teaching?

TABLE 15.4.

REFLECTION GUIDE FOR STUDENTS WITH DISABILITIES

Reflection Guide for Students With Disabilities		
Three-dimensional learning	Effective strategies	Science achievement and education policy
Guiding questions		
How did the strategy employed in the vignette create engagement of the focus group?		
By applying three-dimensional learning and effective strategies to your setting, what implications do you see? What examples can you provide from your classroom or school?		
What questions does the vignette raise?		

1. What types of models did you find in the vignette? How did students develop models? How did students use models?

2. What was the understanding that students gained from the 3-D model with the golf balls? How did the use of this model support deeper understanding? What other models in the vignette were particularly effective in reinforcing similar disciplinary core ideas?

3. How did the crosscutting concept of Patterns contribute to student understanding of the disciplinary core ideas and/or science and engineering practices?

4. What effective strategies did you notice? How did the students with disabilities respond to these strategies? Were there some strategies that you have tried with students with disabilities and would like to use more often in your classroom?

Case Study Utility for Classroom Teaching and Professional Development

TABLE 15.5.

REFLECTION GUIDE FOR ENGLISH LANGUAGE LEARNERS

Reflection Guide for English Language Learners		
Three-dimensional learning	Effective strategies	Science achievement and education policy
Guiding questions		
How did the strategy employed in the vignette create engagement of the focus group?		
By applying three-dimensional learning and effective strategies to your setting, what implications do you see? What examples can you provide from your classroom or school?		
What questions does the vignette raise?		

1. How many science and engineering practices did you find in the vignette? Which practices did you find effective in the vignette? Why did you think that they were effective in this context?

2. What disciplinary core idea did you feel that the students mastered in the vignette? When you trace the development of that idea, how was it strengthened by the teacher? Were there one or two ideas that needed to be further developed to deepen student understanding?

3. How did the crosscutting concept of Stability and Change contribute to student understanding of the disciplinary core ideas and science and engineering practices?

4. What were some of the different strategies you noticed? Were there some strategies that were unfamiliar to you that you would like to try? How did the strategies support both content understanding and language development for the students?

TABLE 15.6.
REFLECTION GUIDE FOR GIRLS

Reflection Guide for Girls		
Three-dimensional learning	Effective strategies	Science achievement and education policy
Guiding questions		
How did the strategy employed in the vignette create engagement of the focus group?		
By applying three-dimensional learning and effective strategies to your setting, what implications do you see? What examples can you provide from your classroom or school?		
What questions does the vignette raise?		

1. How did you see engineering practices develop across the vignette? What was one engineering practice that you found well exemplified in the vignette? What engineering practice that you have not tried in your own classroom would you would like to explore?

2. How did you notice the disciplinary core idea of Engineering develop through the vignette? How did the teacher engage the students as they defined the problem? Were the disciplinary core ideas of Life Sciences addressed simultaneously through the engineering context? How?

3. How did the crosscutting concept of Defining and Delimiting Engineering Problems contribute to student understanding of the disciplinary core ideas and science and engineering practices?

4. What strategies for girls did you notice taking place in the vignette? Which ones were new to you? Thinking about your own teaching and the girls in your classroom, could some of them become more engaged if you employed a few of the strategies highlighted in the vignette?

Case Study Utility for Classroom Teaching and Professional Development

TABLE 15.7.

REFLECTION GUIDE FOR STUDENTS IN ALTERNATIVE EDUCATION

Reflection Guide for Students in Alternative Education		
Three-dimensional learning	Effective strategies	Science achievement and education policy
Guiding questions		
How did the strategy employed in the vignette create engagement of the focus group?		
By applying three-dimensional learning and effective strategies to your setting, what implications do you see? What examples can you provide from your classroom or school?		
What questions does the vignette raise?		

1. This vignette highlighted a number of different practices. How were they supported by comments, prompts, or questions from the teacher? Was there any practice that could have been explored further? How?

2. What disciplinary core idea did the students master by the end of the vignette? What would be your next move with these students in terms of more complete understanding of disciplinary core ideas? How did this vignette portray abstract disciplinary core ideas in a concrete way? Why do you think the teacher made that pedagogical choice?

3. How did the crosscutting concept of Patterns contribute to student understanding of the disciplinary core ideas and science and engineering practices?

4. What was one strategy that you noted in the vignette? How was it used to support student engagement? Was there any strategy that resonated with you in terms of the students and other connections to your own classroom?

TABLE 15.8.
REFLECTION GUIDE FOR GIFTED AND TALENTED STUDENTS

Reflection Guide for Gifted and Talented Students		
Three-dimensional learning	Effective strategies	Science achievement and education policy
Guiding questions		
How did the strategy employed in the vignette create engagement of the focus group?		
By applying three-dimensional learning and effective strategies to your setting, what implications do you see? What examples can you provide from your classroom or school?		
What questions does the vignette raise?		

1. How did the teacher support students to ask questions throughout the vignette? Were the students provided opportunities to become more skilled in the science and engineering practice of Asking Questions? How did you see the practice of Asking Questions lead students to predict causal relationships?

2. As you review the many different experiences with the disciplinary core ideas in the vignette, ask: How did the students become increasingly more sophisticated in their application of these ideas? How did the students own their own growth in understanding disciplinary core ideas throughout the vignette?

3. How did the crosscutting concept of Systems and System Models contribute to student understanding of the disciplinary core ideas and science and engineering practices?

4. What were some of the strategies that you noted and how were they effective for the students? What was one strategy that you would like to use with your gifted and talented students? Why did you pick that one?

INDEX

Page numbers printed in **boldface** *type refer to figures or tables.*

A

A Framework for K–12 Science Education: Practices, Crosscutting Concepts, and Core Ideas, xi, xii, 1, 3, 7, 24, 172, 179
    development of, 23, 25
    false dichotomies related to equity issues, 10–11
    goal of science, 172
    integrating three dimensions of, xi, xii, 30, 181, 187 (*See also* Three-dimensional learning)
    vision of science for all students, 9–10, 29, 35
Accommodations and modifications, 186
    for English language learners, 116
    for students with disabilities, 38, **39,** 83, 84, 94, 96, 97
Achieve Inc., 21, 24, 25, 183
Achievement in science, 2
    of alternative education students, 154
    barriers to, 3, 14, 26, 29, 84, 94, 102, 115, 168
    of economically disadvantaged students, 43, 59
    of English language learners, 101, 116
    foundational capacity development and, 12
    gaps in, xi, 12, 29, 31
    of gifted and talented students, 157, 168
    of girls, 119, 134, 136
    professional development and utility of case study data on, 194
    Reflection Guides for evaluation of, **195–202,** 196–202
    of students from racial and ethnic groups, 61, 80
    of students with disabilities, 83, 96–97
Adequate yearly progress (AYP)
    of English language learners, 116–117
    of students from racial and ethnic groups, 80–81
    of students with disabilities, 98
Advancement Via Individual Determination (AVID), 64–65
After-school opportunities
    for alternative education students, **39,** 139, 143, **145,** 152
    for girls, 120, 127
    for students from racial and ethnic groups, 62
Alternative education students, xiv, 4, 33, **34,** 139–155
    demographics of, 139–140, 154
    education policy for, 139, 154–155
    effective strategies for, **39,** 152–154
    science achievement of, 154
    types of schools for, 152, 154
    vignette: constructing explanations about energy in chemical processes, 139–152
        alternative education connections, 140–147
        CCSS connections, 151–152, **153**
        developing explanations for chemical properties of matter, 143–146, **145**
        finding patterns in the periodic table, 142–143
        introducing career connections to chemistry, 140–142, 147
        introduction, 140
        NGSS connections, 147–151, **153**
        opportunity for discourse, 186
        Reflection Guide, 201, **201**
        using evidence to develop claims, 146–147
Alternative energy: vignette of students from racial and ethnic groups, 61–78
America Diploma Project (ADP), 21
American Association for the Advancement of Science, 1
American Community Survey, 58
American Recovery and Reinvestment Act of 2009 (ARRA), 22, 24
Analysis and reasoning capacity, 11, 12, 14–15
Animal structure and function: vignette of gifted and talented students, 157–165
Appendix D: All Standards, All Students, 4–5, 27, 29, **30,** 30–32, 37
    context of science learning, 31–32
    effective classroom strategies for teachers, 31, **31**
    opportunities and challenges for all students, 30–31
Appendix F: Science and Engineering Practices, 32–33
Appendix G: Crosscutting Concepts, 33
Appendix H: Understanding the Scientific Enterprise: The Nature of Science, 33
Appendix I: Engineering Design, 33
Appendix J: Science, Technology, Society and the Environment, 33
Argumentation, 11, 14, 15, 31, 33, 174
    in vignette of economically disadvantaged students, 49–50, 53, 54, 55, **57,** 184
    in vignette of English language learners, 103, 104, 113
    in vignette of gifted and talented students, 157–165, **167**
    in vignette of girls, 126, 129, 130, 131, 132, **135**
    in vignette of students from racial and ethnic groups, 68, 72, **72,** 74, 76, **79,** 197
Assessments, 26
    of English proficiency, 116–117
    international, 2
    National Assessment of Educational Progress, 31
        for alternative education students, 154
        for economically disadvantaged students, 43, 59
        for English language learners, 116
        for gifted and talented students, 168
        for girls, 119, 136
        for students from racial and ethnic groups, 61, 80
        for students with disabilities, 96–97

INDEX

performance of gifted and talented students on, 168
standards-based accountability testing, 12
for students with disabilities, 83, 91
At-risk students, 10, 11, 98, 139, 140, 152, 154. *See also* Alternative education students

B
Barriers to learning, 29, 84, 94, 168
language, 3, 14, 26, 102, 115
Benchmarking for Success: Ensuring U.S. Students Receive a World-Class Education, 24
Bias reviews of the *NGSS,* xi, 4, 29, **30,** 32
Bilingual education, 40, 116–117
Bilingual students, 26, 115, 171. *See also* English language learners

C
Carbon cycle, 62–63, 65, 67, 68, 70, 73
Career and college readiness, xi, 5, 8, 22, 26, 27, 33, 35, 172
Careers in STEM fields, 1, 2, 8, 30, 78, 80, 81, 183, **190**
for alternative education students, 140–142, 147
for girls, 119, 132, 134, 136–137
for students from racial and ethnic groups, 185
Carnegie Corporation, 23, 24, 25, 26
Carnegie/Institute for Advanced Study Commission, 23
Case studies, xi–xiii, 4–5, 9, 11, 29, 33–35, **34**
alternative education students, 139–155
caveats to understand purpose of, 38–40
economically disadvantaged students, 43–59
English language learners, 101–117
gifted and talented students, 157–169
girls, 119–137
how to approach vignettes in, 40–42
students from racial and ethnic groups, 61–81
students with disabilities, 83–98
synthesis of effective classroom strategies across, 37–38
use by collaborative learning communities, 37
using to inform unit design, 37, 171–177
utility for classroom teaching and professional development, 193–202
Chemical processes
constructing explanations about energy in: vignette of alternative education students, 139–152
developing conceptual models to explain: vignette of economically disadvantaged students, 43–56
Classroom Opportunities Multiply with Practices and Application of Science Standards (COMPASS), 25
Clinton, Bill, 22
Common Core of Data report, 58
Common Core State Standards in English Language Arts *(CCSS ELA)* and Mathematics *(CCSS Mathematics),* xi, xii, 16, 21–24
NGSS connections to, 30–31, 37
in vignette of alternative education students, 151–152, **153**
in vignette of economically disadvantaged students, 55–56, **57**
in vignette of English language learners, 42, 112–113, **114**
in vignette of gifted and talented students, 165, **167**
in vignette of girls, 133, **135**
in vignette of students from racial and ethnic groups, 77–78, **79**
in vignette of students with disabilities, 93–94, **95**
Community involvement, with students from racial and ethnic groups, **39,** 61, 72, 78
Compacting curriculum, 159, 166
Computational thinking, 14, 85, 103. *See also* Mathematics
Council of Chief State School Officers (CCSSO), 21, 24, 25
Council of State Science Supervisors (CSSS), 25
Crosscutting concepts, xi, xii, 9, 10, 30, 33, 179, 181, 187
Reflection Guides to evaluate learning of, 196–202
in vignette of alternative education students, 151, **153**
in vignette of economically disadvantaged students, 54, **57,** 184
in vignette of English language learners, 112, **114,** 184
in vignette of gifted and talented students, 165, **167**
in vignette of girls, 133, **135**
in vignette of students from racial and ethnic groups, 77, **79**
in vignette of students with disabilities, 93, **95,** 184
Culturally relevant teaching, 17–18, 31, 37, 38, 40, 41, 176
for economically disadvantaged students, 43, 48, 50, 56
for English language learners, 106, 115, 117
for girls, 185
for students from racial and ethnic groups, **39,** 61, 64, 73, 78

D
Developing Assessments for the Next Generation Science Standards, 25
Differentiated instruction, 186, **192**
for gifted and talented students, 157, 159, 166
for students with disabilities, **39,** 94
Disabilities, students with, xiii, 4, 25, 33, **34,** 83–98
accommodations and modifications for, 38, **39,** 83, 84, 94, 96, 97
adequate yearly progress of, 98
demographics of, 83, 96
differentiated instruction for, **39,** 94
education policy for, 97–98
effective strategies for, 38, **39,** 83, 94
Individualized Education Plans for, 32, 83, 84, 90, 94, 96, 97
science achievement of, 83, 96–97
Specific Learning Disabilities, 98
vignette: using models of space systems to describe patterns, 83–94
assessing student learning, 83, 91
CCSS connections, 93–94, **95**
exploring Earth-Sun-Moon relationship, 85–87
exploring Moon phases, 87–91, **89, 90**
introduction, 84
NGSS connections, 91–93, **95**
Reflection Guide, 198, **198**
special education connections, 84–91

INDEX

three-dimensional learning, 184
Disciplinary core ideas, xi, xii, 9, 30, 174, 179, 181, 187
    Reflection Guides to evaluate learning of, 196–202
    in vignette of alternative education students, 149, **153**
    in vignette of economically disadvantaged students, 52, **57**, 184
    in vignette of English language learners, 110, **114**, 184
    in vignette of gifted and talented students, 163, **167**
    in vignette of girls, 131, **135**
    in vignette of students from racial and ethnic groups, 75, **79**
    in vignette of students with disabilities, 92, **95**, 184
Discourse strategies. *See* Science discourse
Diverse student groups, xiii, 4, 25, 33
    alternative education students, 139–155
    case studies in *NGSS* of, xi–xiii, 4–5, 9, 11, 29, 33–35, **34** (*See also* Case studies)
    economically disadvantaged students, 43–59
    effective classroom strategies across, 38, **39**
    English language learners, 101–117
    gifted and talented students, 157–169
    girls, 119–137
    good teaching for, 40
    implementing *NGSS* with, xi–xiii
    importance of three-dimensional learning for, 182–183
    professional development and classroom teaching of, 193–202, **195–202**
    research literature on, 37–38
    science education access for, 26
    sensitivity to, 38
    students from racial and ethnic groups, 61–81
    students with disabilities, 83–98
    unit design for three-dimensional learning of, 171–177
        application of three shifts, 176–177
        building on place-based context, 175–176
        driving questions, 173–174
        engaging in phenomena, 171–173
    variability of students within, 39
Diversity and Equity Team, xi, 3–4, 9, 29–35, **30**, 37, 40
Driving questions, 171, 173–174, 176, 177, 180
    importance for diverse student groups, 174
    importance for *NGSS*, 174
    in vignette of alternative education students, 142
    in vignette of economically disadvantaged students, 45, 58
    in vignette of English language learners, 173
    in vignette of students from racial and ethnic groups, 65, 77
Dropout prevention schools, 139, 152, 154, 155

E

Earth's surface systems: vignette of English language learners, 101–113
Economically disadvantaged students, xiii, 4, 10, 12, 25, 33, **34**, 43–59
    in alternative education, 139
    demographics of, 43, 58–59
    education policy for, 59

    effective strategies for, **39**, 43, 44, 56–58
    gifted and talented, 168
    science achievement of, 43, 59
    vignette: developing conceptual models to explain chemical processes, 43–56
        applying scientific knowledge to an engineering problem, 50–51
        CCSS connections, 55–56, **57**
        developing initial conceptual model, 45–47, **47**
        economically disadvantaged connections, 44–51
        gathering new evidence to evaluate and revise conceptual models, 47–49, **48**
        introduction, 44
        NGSS connections, 51–54, **57**
        opportunity for discourse, 185–186
        Reflection Guide, 196, **196**
        three-dimensional learning, 184
        using literacy, discourse, and argumentation to develop shared understanding, 49–50
Ecosystems
    constructing explanations to compare cycle of matter and flow of energy through local ecosystems: vignette of students from racial and ethnic groups, 61–78
    defining problems with multiple solutions within an ecosystem: vignette of girls, 119–133
"Educate to Innovate" campaign, 137
Education for All Handicapped Children Act (Public Law 94-142), 96, 97
Education policy
    for alternative education students, 139, 154–155
    for economically disadvantaged students, 59
    for English language learners, 116–117
    for gifted and talented students, 168–169
    for girls, 136–137
    professional development and utility of case study data on, 194
    Reflection Guides for evaluation of, **195–202**, 196–202
    for students from racial and ethnic groups, 80–81
    for students with disabilities, 97–98
Educators Evaluating the Quality of Instructional Products (EQuIP) Rubric, 183
Effective classroom strategies, xi–xii
    for alternative education students, **39**, 152–154, **39**, 152–154
    in Appendix D, 31, **31**
    for economically disadvantaged students, **39**, 43, 44, 56–58
    for English language learners, 38, **39**, 40, 101, 113–115
    for gifted and talented students, **39**, 157, 166
    for girls, **39**, 119, 133–134
    professional development and utility of case studies for, 194
    Reflection Guides for evaluation of, **195–202**, 196–202
    for students from racial and ethnic groups, **39**, 61, 78–80
    for students with disabilities, 38, **39**, 83, 94
    synthesis across case studies, 37–38, **39**
    three-dimensional learning blended with, 40
Elementary and Secondary Education Act (ESEA), 59, 80, 83, 98, 116, 154, 168–169

INDEX

Emotional and social capacity, 11, 12, 16–17
Engineering design, 15, 17. *See also* Science and engineering practices
    diversity and equity topic in Appendix I on, 33
    in vignette of economically disadvantaged students, 50–51, 52, **57**
    in vignette of English language learners, 107–108, **108**, 109
    in vignette of girls, 119–133, **135**
English for Speakers of Other Languages (ESOL), 113
English Language Acquisition, Language Enhancement, and Academic Achievement Act, 116
English language arts, *NGSS* connections to *CCSS* for
    in vignette of alternative education students, 151, **153**
    in vignette of economically disadvantaged students, 55, **57**
    in vignette of English language learners, 112–113, **114**
    in vignette of gifted and talented students, 165, **167**
    in vignette of girls, 133, **135**
    in vignette of students from racial and ethnic groups, 77, **79**
    in vignette of students with disabilities, 93, **95**
English language learners (ELLs), xiii, 4, 25, 26, 33, **34**, 101–117
    adequate yearly progress of, 116, 118
    in alternative education, 139
    bias review of the *NGSS* for, 32
    demographics of, 101, 115–116
    education policy for, 116–117
    effective strategies for, 38, **39**, 40, 101, 113–115
    "English only" policy for, 116
    heterogeneity of, 39, 102, 115
    science achievement of, 101, 116
    vignette: developing and using models to represent Earth's surface systems, 40–41, 101–113
        attention to context, 106, 185
        CCSS connections, 42, 112–113, **114**
        ELL connections, 101–108, **102, 103, 108**
        introduction, 101
        NGSS connections, 41, 109–112, **114**
        opportunity for discourse, 186
        Reflection Guide, 199, **199**
        three-dimensional learning, 184
Equal Access to Language and Science (EquALS) Rubric, xii, 37, 179, 183–187
    criterion 1: focus on three-dimensional learning, 183–184, 189
    criterion 2: attention to context, 184–185, 190
    criterion 3: opportunity for discourse, 185–186, 191
    criterion 4: emphasis on student thinking and reflection, 186–187, 192
Equitable learning opportunities, 31, **31**

F
Family outreach, **39**, 139, 152
Fordham Institute, 25
Forest restoration project: vignette of girls, 119–133
Foundation box, xii

for vignette of alternative education students, **153**
for vignette of economically disadvantaged students, **57**
for vignette of English language learners, **114**
for vignette of gifted and talented students, **167**
for vignette of girls, **135**
for vignette of students from racial and ethnic groups, **79**
for vignette of students with disabilities, **95**
Foundational capacity development, 11–17
    analysis and reasoning, 14–15
    emotional and social capacity, 16–17
    language, 13–14
    representation and symbolization, 15–16
Fox, L., 40
Framework. *See A Framework for K–12 Science Education: Practices, Crosscutting Concepts, and Core Ideas*

G
Gates, James, 2
Gifted and talented students, xiv, 4, 33, **34**, 157–169
    definitions of, 167–168, 169
    demographics of, 157, 167–168
    differentiated instruction for, 157, 159, 166
    education policy for, 168–169
    effective strategies for, **39**, 157, 166
    funding of programs for, 168, 169
    science achievement of, 157, 168
    vignette: constructing arguments about interaction of structure and function in plants and animals, 157–165
        CCSS connections, 165, **167**
        emphasis on student thinking and reflection, 187
        gifted and talented connections, 159–162, **160, 161**
        introduction, 158
        NGSS connections, 162–165, **167**
        Reflection Guide, 202, **202**
Girls, xiv, 4, 10, 25, 33, **34**, 119–137
    demographics of, 136
    education policy for, 136–137
    effective strategies for, **39**, 119, 133–134
    science achievement of, 119, 134, 136
    vignette: defining problems with multiple solutions within an ecosystem, 119–133
        analyzing and interpreting data, 124–125
        attention to context, 185
        CCSS connections, 133, **135**
        connections to girls in science and engineering, 121–129
        emphasis on student thinking and reflection, 186
        engaging in phenomena, 171–172
        engineering: the solution, **121**, 121–122
        introduction, 120
        multiple design solutions to an engineering problem, 126–129
        NGSS connections, 129–133, **135**
        planning an investigation, 122–124, **124**
        Reflection Guide, 200, **200**
        using modeling to define and refine an engineering problem, **125**, 125–126, **126**
Goal of science, 172

INDEX

Goals 2000: Education America Act, 22
Goals for science for all students, 1, 3–5, 7–11

H

Habitats for forest wildlife: vignette of girls, 119–133
Home culture connections, **39**, 101, 106, 113, 115
Home language support, **39**, 101, 105, 113, 115
Hunt Institute, 25

I

IDEA Partnership, 98
Individualized academic support, **39**, 139, 142, 152, 154
Individualized Education Plans (IEPs), 32, 83, 84, 90, 94, 96, 97
Individuals with Disabilities Education Act (IDEA), 96, 97–98
Inquiry-based teaching, 2, 23, 33, 56, 120, 134, 187

J

Jacob Javits Gifted and Talented Students Education Act, 168–169

L

Ladson-Billings, G., 40
Language barriers, 3, 14, 26, 102, 115
Language processing difficulties, 32–33
Language skills, 11, 12, 13–14
Learning progressions, 34, 40, 51, 91, 156
Life Cycle Assessment Process Tool, 67, 70
Life skills training, 139, 140, 152
Limited English proficient (LEP) students, 115. *See also* English language learners
Literacy for Science: Exploring the Intersection of the Next Generation Science Standards and Common Core for ELA Standards, a Workshop Summary, 14
Literacy skills, 2, 11, 12, 13–14, **39**, 73, 78, **191**. *See also* English language arts
   for English language learners, 101–117, 173
   gaps between states, 26
   reading, 2, 14, 15, 26, 31, 40, 42, 49, 55, 81, 84, 90–91, 93, 113, 151, 158, 159, 165
   writing, 2, 14, 21, 26, 48, 55, 77, 90, 93, **102**, 105, 107, 111, 113, 165, 175
Loeb, S., 40

M

Mathematics, 2, 12, 13, 14, 23, 37
   NGSS connections to *CCSS* for, 30–31, 37
      in vignette of alternative education students, 152, **153**
      in vignette of economically disadvantaged students, 55–56, **57**
      in vignette of English language learners, 112–113, **114**
      in vignette of gifted and talented students, 165, **167**
      in vignette of girls, 133, **135**
      in vignette of students from racial and ethnic groups, 77–78, **79**
      in vignette of students with disabilities, 94, **95**
Mentors
   for alternative education students, 142, 154
   for girls, 134
   for students from racial and ethnic groups, **39**, 61, 78
Metacognition, 134, 186, **192**
Mindset, 11, 17
Models, development and use of, 9, 10, 11, 13, 14–16, 17
   professional development and, 193, 196, 198
   three-dimensional learning and, 180, 181, 193
   in vignette of alternative education students, 142, 148, 150, **153**
   in vignette of economically disadvantaged students, 43–59, **47, 48, 57**, 184, 185
   in vignette of English language learners, 101–113, **103, 114**, 173, 184
   in vignette of gifted and talented students, 159–160, 164, 165, **167**
   in vignette of girls, 120, 125, 127, 129
   in vignette of students from racial and ethnic groups, 65–67, 70–72, **71**, 76, **79**
   in vignette of students with disabilities, 86–93, **89, 95**, 184
The Moon, 90, 93
Moon phases: vignette of students with disabilities, 83–94
Motivation of students, 11, 13, 17, 32, 33, 42, 174, 175, 194
   gifted and talented, 166, 187
   girls, 127, 134
   racial and ethnic minorities, 62, 78
Multimodal experiences, 14, 179, **192**, 187
   for students from racial and ethnic groups, **39**, 61, 67, 78
Multiple means of action and expression, 83, 89, 94
Multiple means of engagement, 83, 91, 94, 154
Multiple modes of representation, 14, 16, 38, 63, 78
   for English language learners, 113–115
   for students with disabilities, 83, 84, 86–87, 90, 94

N

National Academies of Sciences, Board on Science Education, 23
National Assessment of Educational Progress (NAEP), 31
   for alternative education students, 154
   for economically disadvantaged students, 43, 59
   for English language learners, 116
   for gifted and talented students, 168
   for girls, 119, 136
   for students from racial and ethnic groups, 61, 80
   for students with disabilities, 96–97
National Association for Gifted Children, 167
National Association of State Boards of Education (NASBE), 25
National Governors Association Center for Best Practices (NGAC), 21, 24
National Research Council (NRC), 1, 25, 179
National School Lunch Program (NSLP), 58, 59
National Science Teachers Association (NSTA), 1, 25, 183
Nation's Report Card: Science 2009, 59
New England Common Assessment Program (NECAP), 21
Next Generation Science Standards (NGSS), 1–5
   Appendix D: All Standards, All Students, 4–5, 27, 29, **30**, 30–32, 37

INDEX

Appendix F: Science and Engineering Practices, 32–33
Appendix G: Crosscutting Concepts, 33
Appendix H: Understanding the Scientific Enterprise: The Nature of Science, 33
Appendix I: Engineering Design, 33
Appendix J: Science, Technology, Society and the Environment, 33
bias reviews of, xi, 4, 29, **30,** 32
building policy support for, 21–27
case studies of diverse student learners and, xi–xiii, 4–5, 9, 11, 29, 33–35, **34**
    alternative education students, 139–155, **153**
    economically disadvantaged students, 43–59, **57**
    English language learners, 101–117, **114**
    gifted and talented students, 157–169, **167**
    girls, 119–137, **135**
    students from racial and ethnic groups, 61–81, **79**
    students with disabilities, 83–98, **95**
Common Core State Standards and, 21–24, 27
connections to, 30–31, 37
conceptual framework guiding diversity and equity in, 37–42
development of, 1, 3–5, 24
Diversity and Equity Team for, xi, 3–4, 9, 29–35, **30,** 37, 40
diversity and equity topic in Appendixes for, 32–33
false dichotomies related to equity issues, 10–11
implementing with diverse student groups, xi–xiii
importance of phenomena for, 172
lack of federal support for, 24
launching of, 23–26
lead state teams for, 24
partnerships for, 24–25
performance expectations for (*See* Performance expectations)
reflecting on instruction to promote equity and alignment to, 179–192 (*See also* Equal Access to Language and Science (EquALS) Rubric)
review process for, 29
three-dimensional learning and, xi, xii, 30, 34, 179–187 (*See also* Three-dimensional learning)
vision of science for all students, 2, 3–5, 7–11, 29, 35
NGSS Volume 2: Appendixes, 32
No Child Left Behind Act of 2001 (NCLB), 32, 58, 154
Numeracy. *See* Mathematics

O

Obama, Barack, 137
Opportunity Equation: Transforming Mathematics and Science Education for Citizenship and the Global Economy, 23

P

Performance expectations, xi, 39, 41, 174, 176
    bias review of language used in, 32
    clarification statements in, 32
    for three-dimensional learning, 181
    in vignette of alternative education students, 139, 148, **153**
    in vignette of economically disadvantaged students, 43–44, 52, **57**
    in vignette of English language learners, 101, 109, **114,** 173
    in vignette of gifted and talented students, 157–158, 163, **167**
    in vignette of girls, 119, 120, 129–130, 132, **135**
    in vignette of students from racial and ethnic groups, 61–62, 73–74, **79**
    in vignette of students with disabilities, 83, 91–92, 94, **95**
Periodic table, 140, 141, 142, 146, 148, 149, 151, **153**
Persistence, 10, 11, 12, 17
Phenology wheel, **121,** 121, **125,** 125–126, **126,** 127, 128, 132, 187
Phenomena, engaging in, 9, 10–11, 13, 15–16, 17, 30, 35, 171–173, 176–177, 179
    discourse for, **191**
    driving questions for, 174, 176
    importance for diverse student groups, 173
    importance for *NGSS,* 172
    place-based context for, 175, 176, 177, **190**
    three-dimensional learning and, 179, 181, 182, 183–184, 185, 187, **189**
    in vignette of alternative education students, 150, 151
    in vignette of economically disadvantaged students, 46, 47, 53, 54, 55, **57**
    in vignette of English language learners, 111, 113
    in vignette of gifted and talented students, 164
    in vignette of girls, 132, 134, 171–172
    in vignette of students with disabilities, 92, 93, **95**
Place-based context, building on, 41, 171, 175–176
    importance for diverse student groups, 175–176
    importance for *NGSS,* 175
    for three-dimensional learning, 184–185
    in vignette of English language learners, 106, 185
    in vignette of girls, 185
    in vignette of students from racial and ethnic groups, 64, 78, 175, 185
Plant structure and function: vignette of gifted and talented students, 157–165
Poverty, 43, 58–59, 102. *See also* Economically disadvantaged students
Professional development, xii, xiii, 25
    utility of case studies for classroom teaching and, 193–202
    instruction with effective strategies, 194
    instruction with three-dimensional learning, 193–194
    Reflection Guides, 194, **195–202,** 196–202
    science achievement data and education policy, 194
Project-based learning, **39,** 43, 46, 56–58
Public Law 94-142 (Education for All Handicapped Children Act), 96, 97
Public Law 103-239 (School to Work Opportunities Act of 1994), 136
"Push-in" schools, 154

Q

Quinn, Helen, 23

INDEX

R
Race to the Top, 22, 24, 81
Racial and ethnic groups, students from, xiii, 4, 10, 25, 33, **34,** 61–81
    adequate yearly progress of, 80–81
    in alternative education, 139
    demographics of, 61, 80
    education policy for, 80–81
    effective strategies for, **39,** 61, 78–80
    science achievement of, 61, 80
    vignette: constructing explanations to compare cycle of matter and flow of energy through local ecosystems, 61–78
        application of core ideas, **71,** 71–73, **72**
        attention to context, 64, 78, 175, 185
        building on students' background knowledge, 62–64
        CCSS connections, 77–78
        comparing cycle of matter to flow of energy in carbon cycle, 70–71
        definition of efficiency, 64–66
        emphasis on student thinking and reflection, 186
        introduction, 62
        NGSS connections, 73–77
        racial and ethnic connections, 62–73
        Reflection Guide, 197, **197**
        revising model and constructing explanations, 66–70, **69**
Reading skills and tasks, 2, 14, 15, 26, 31, 40, 42, 49, 55, 81, 84, 90–91, 93, 113, 151, 158, 159, 165, **191**
Ready, Set, Science! Putting Research to Work in K–8 Science Classrooms, xii
Reflection Guide, 194
    for alternative education students, 201, **201**
    for economically disadvantaged students, 196, **196**
    for English language learners, 199, **199**
    general, **195**
    for gifted and talented students, 202, **202**
    for girls, 200, **200**
    for racial and ethnic groups, 197, **197**
    for students with disabilities, 198, **198**
Rehabilitation Act of 1973, Section 504, 96, 97
Representation and symbolization capabilities, 11, 12, 15–16
Response to intervention (RTI) model
    for gifted and talented students, 168
    for students with disabilities, 98
Richard B. Russell National School Lunch Act, 58
Role models, **39,** 61, 78, 185, **190**

S
Safe learning environment, **39,** 139, 152, 154
School support systems, 38, **39**
    for students from racial and ethnic groups, 61, 64–65, 78–80
School to Work Opportunities Act of 1994 (Public Law 103-239), 136
Science, technology, engineering, and mathematics (STEM) careers, 1, 2, 8, 30, 78, 80, 81, 183, **190**
    for alternative education students, 140–142, 147
    for girls, 119, 132, 134, 136–137
    for students from racial and ethnic groups, 185
Science, technology, engineering, and mathematics (STEM) education, 2, 5, 81, 136, 137, 194
    steps toward improving for underserved students, 183–187, 189–192 (*See also* Equal Access to Language and Science (EquALS) Rubric)
Science and engineering practices, xi, xii, 2, 7–18, 30, 179, 181, 187
    diversity and equity topic in Appendix F on, 32–33
    equity and goals of science for all students, 7–8
    false dichotomies related to equity issues, 10–11
    foundational capacity development through, 11–17
        analysis and reasoning, 14–15
        emotional and social capacity, 16–17
        language, 13–14
        representation and symbolization, 15–16
    Reflection Guides to evaluate learning of, 196–202
    student engagement in, 10–11
    in vignette of alternative education students, 150, **153**
    in vignette of economically disadvantaged students, 53–54, **57,** 184
    in vignette of English language learners, 111, **114,** 184
    in vignette of gifted and talented students, 164, **167**
    in vignette of girls, 132–133, **135**
    in vignette of students from racial and ethnic groups, 76, **79**
    in vignette of students with disabilities, 91, 92–93, **95,** 184
    and vision of science for all students, 9–10
Science discourse, 13–14, 17, 32, 34, 35, 49, 174, 176
    for English language learners, 101, 104, 113–115
    teaching to support opportunity for, 183, 185–186, 187, **191**
Science for all students, 2, 3–5, 7–11, 29, 35
Scientific literacy, 1, 2, 5, 7, 62, 172, 175
Scientific thought processes, 2
Section 504 of Rehabilitation Act of 1973, 96, 97
Self-direction, **39,** 157, 162, 166
Self-efficacy, 12, 154, 187
Self-regulation, 11, 12, 17
Simon, Seymour, 90
Social activism, 61, 72, 78–80
Social Security Act, Title IV, 58
Soil profiles: vignette of English language learners, 101–113
Soland, J., 40
Space systems: vignette of students with disabilities, 83–94
Special education, 44, 83–98. *See also* Disabilities, students with
Specific Learning Disabilities (SLDs), 98. *See also* Disabilities, students with
Sputnik age, 1–2
The State of State Science Standards 2012, 25
Strategic grouping of students, 84, 157, 158, 166
Student thinking and reflection, 183, 186–187

T
Taking Science To School: Learning and Teaching Science in Grades K8, xii, 9

INDEX

Teachers
    case study utility for classroom teaching and professional development of, 193–202, **195–202**
    effective strategies of, xi–xii, 38, **39,** 40 (*See* Effective classroom strategies)
    fundamental beliefs and practices of, 38
    reflecting on instruction to promote equity and alignment to *NGSS,* 179–192 (*See also* Equal Access to Language and Science (EquALS) Rubric)
Technological society, 2
Three-dimensional learning, xi, xii, 30, 32, 34, 40, 179–184, 187. *See also* Crosscutting concepts; Disciplinary core ideas; Science and engineering practices
    explanation of, 181–182
    focus on, as criterion of Equal Access to Language and Science (EquALS) Rubric, 183–184, 189
    importance for diverse student groups, 182–183
    performance expectations for, 181
    professional development considerations for, 193–194
    Reflection Guides for evaluation of, **195–202,** 196–202
    teaching cases and, 179–181
    unit design and shifts for promotion of, 171–177
        application for teaching diverse student groups, 176–177
        building on place-based context, 175–176
        driving questions, 173–174
        engaging in phenomena, 171–173
    in vignette of economically disadvantaged students, 184
    in vignette of English language learners, 184
    in vignette of students from racial and ethnic groups, 62, 186
    in vignette of students with disabilities, 84, 85, 184
Title I of the Elementary and Secondary Education Act, 59, 81, 98, 169
Title III of the Elementary and Secondary Education Act, 116
Title IV of the Elementary and Secondary Education Act, 168
Title IV of the Social Security Act, 58
Title IX of the Elementary and Secondary Education Act, 136, 169
21st-century skills, 2, 8
Twice-exceptional children, 168

U
Unit design for three-dimensional learning, xii, 37, 171–177
    application of three shifts for teaching diverse student groups, 176–177
    building on place-based context, 175–176
        importance for diverse student groups, 175–176
        importance for *NGSS,* 175
        in vignette of students from racial and ethnic groups, 175
    driving questions, 173–174
        importance for diverse student groups, 174
        importance for *NGSS,* 174
        in vignette of English language learners, 173
    engaging in phenomena, 171–173
        importance for diverse student groups, 173
        importance for *NGSS,* 172
        in vignette of girls, 171–172
Universal Design for Learning, **39,** 89, 90, 91, 94
U.S. Census Bureau, 58, 61, 62, 80
U.S. Office of Special Education, 98
Usable knowledge, 171, 172, 174, 77, 182

V
Vision of science for all students, 2, 3–5, 7–11, 29, 35
Vocabulary development, 2, 14, 105, 115, 124, 179–180, 191. *See also* Language skills; Literacy skills
Vocational programs, 140

W
WIDA Consortium, 116
World Class Instructional Design, 116
Writing skills and tasks, 2, 14, 21, 26, 48, 55, 77, 90, 93, **102,** 105, 107, 111, 113, 165, 175, **191**